# Corals of Florida and the Caribbean

**University Press of Florida**

Florida A&M University, Tallahassee
Florida Atlantic University, Boca Raton
Florida Gulf Coast University, Ft. Myers
Florida International University, Miami
Florida State University, Tallahassee
New College of Florida, Sarasota
University of Central Florida, Orlando
University of Florida, Gainesville
University of North Florida, Jacksonville
University of South Florida, Tampa
University of West Florida, Pensacola

I0103484

**George F. Warner**

# Corals of Florida & the Caribbean

**University Press of Florida**

Gainesville/Tallahassee/Tampa/Boca Raton

Pensacola/Orlando/Miami/Jacksonville/Ft. Myers/Sarasota

17  16  15  14  13  12   6  5  4  3  2  1

*Library of Congress Cataloging-in-Publication Data*
Warner, George F.
Corals of Florida and the Caribbean / George F. Warner.
p. cm.
Includes bibliographical references and index.
ISBN 978-0-8130-4165-0 (alk. paper)—ISBN 0-8130-4165-1 (alk. paper)
    1. Corals—Florida—Identification. 2. Corals—Caribbean Area—
Identification. 3. Corals—Evolution—Caribbean Area. 4. Corals—
Evolution—Florida. I. Title.
QL377.C5W37 2012
593.609729—dc23
2012009928

The University Press of Florida is the scholarly publishing agency for the State
University System of Florida, comprising Florida A&M University, Florida
Atlantic University, Florida Gulf Coast University, Florida International
University, Florida State University, New College of Florida, University of
Central Florida, University of Florida, University of North Florida, University
of South Florida, and University of West Florida.

University Press of Florida
15 Northwest 15th Street
Gainesville, FL 32611-2079
http://www.upf.com

*To all those with whom I explored*
*the coral reefs of the Caribbean*

# Contents

# Preface

The writing of this book resulted from a series of serendipities. My long association with Jamaica, where I learned to dive and saw my first coral reef, was the original lucky break. This was in 1963, before the modern threats to coral reefs had been recognized. Even at that time, however, evidence of overfishing was present, as were instances of pollution, but the impression of the pristine coral reef was fairly clear, and one felt it to be a timeless, stable ecosystem that would endure. More recently, employed in Jamaica at the University of the West Indies, I found the reefs greatly degraded, but not as drastically as had been implied in some of the scientific literature. Many species of coral were still present, and although a few reefs were devoid of significant coral, others were in moderately good health. As part of my work, I measured coral abundance, identified the species, and photographed them. I built up a large collection of images, suitable for publication as a guide to the species of Caribbean corals, but more was needed, since without information on the biology and ecology of corals, an identification guide is little more than a detailed picture book. To understand the threats to modern coral reefs and the dynamic balance of this ecosystem, knowledge of the biology and ecology of corals is essential. So the book developed around the images, and I hope it will serve not only to identify Caribbean reef corals but also to explain why the reefs are in their present degraded state and why their conservation is so very urgent.

The scientific information in this book is based mostly on

available literature and is uncontroversial; therefore, to avoid encumbering the text with frequent references, I have restricted citations to recent reviews wherever possible. Exceptions occur where recent research is involved, such as the use of molecular genetics to investigate the relatedness of corals, where I refer to the relevant articles. In either case, it is relatively easy to find further information online.

Most of the photographs were taken along the coasts of Jamaica, but I also have experience of other locations including the Bahamas, Cayman, Haiti, the Grenadines, Trinidad and Tobago, Margarita Island, Curaçao, and Panama. Although there are some local differences, the coral communities and most of the species are common to all these places and also to Florida and the Keys. Photographs were taken with Nikonos underwater cameras and with a compact digital camera in a case—the latter being, in my opinion, by far the most convenient system. I am grateful to J. C. Ogden for permission to use his photographs of coral spawning (plates 1.13 and 1.14).

The gradual accumulation of knowledge and experience that make it possible to write a book like this has very many roots. I could not have achieved it without the advice and help of many colleagues, friends, and diving buddies going back several decades and now too numerous to list. In addition, the University of the West Indies (UWI) in Jamaica and the University of Trinidad and Tobago (UTT) were essential to the task, since by employing me they made the directly related field work possible. Particularly important were the facilities at UWI's marine laboratories at Discovery Bay and Port Royal on the north and south coasts of Jamaica, respectively, and the Institute of Marine Affairs in Trinidad. To the staff of all three laboratories and associated departments I owe very special thanks. I also thank two reviewers for their positive remarks on the manuscript and more importantly for their valuable criticisms, which I hope I have been able to address successfully. Notwithstanding, any errors that remain are my own.

# Introduction

This book originates from the apparent need, at a critical time for coral reefs, to bring together important aspects of the life of Caribbean corals in a single, easily portable volume. The largest part of the book gives the details required to identify Caribbean reef-building corals to species level. Other chapters explain the basics of their biology and ecology, the modern environmental threats that face them, and the conservation issues that challenge coral reef managers. A final chapter provides a brief history of corals and reefs. The book is therefore intended for all those interested in looking at, studying, using, or protecting Caribbean coral reefs, and for those who would like both themselves and their descendants to continue to enjoy them in the future.

In comparison to the Indo-Pacific region, the Caribbean is a microcosm of the coral universe. It is a much smaller area with a fraction (approximately one-tenth) of the number of species found in the Indo-Pacific. Despite its relatively small extent, however, the Caribbean presents a wide range of environmental conditions in which virtually all aspects of coral diversity and threats to corals may be found. Such species diversity as does exist is not as bewildering as it might appear in the Red Sea, or around the Indonesian archipelago, and much important scientific research on corals and coral reefs has taken place in the Caribbean. In addition, the coastal margins of the Caribbean and its islands are often densely populated. Coastal tourism is an important economic activity, with tourists coming from many parts of the Northern

Hemisphere. The beauty, utility, and health of Caribbean corals ought therefore to be of interest to many people within and well beyond the borders of the Caribbean.

The borders of the Caribbean Sea are relatively easy to define, but to include all Caribbean corals, the Gulf of Mexico, Florida, and the Bahamas need also to be included. Thus the southern, western, and northern boundaries of this wider Caribbean region are defined by the continental land masses of South, Central, and North America, while the eastern boundary is the curve of islands from Trinidad in the south, through the Lesser and Greater Antilles, the Turks and Caicos Islands, and the Bahamas. The eastern coast of Florida is also Caribbean in the character of its corals, and the same is true of Bermuda, but in both cases, lower winter temperatures have resulted in reduced richness of coral species. To the southeast, the huge outflow of fresh water and silt from the Orinoco and Amazon rivers creates a natural boundary to the southward dispersal of corals. A somewhat different and less diverse coral fauna exists further south, off Brazil.

The book is organized in what, I hope, is a logical fashion, starting with the question "what is a coral?" and going on to discuss aspects of the biology of corals. Then the Caribbean coral families and species are described, and their appearance illustrated by photographs. Within each species description I have tried to indicate typical habitat and possible identification difficulties. Caribbean coral reef ecology is then addressed with particular reference to the corals themselves rather than to the vast number of other animals and plants that live on coral reefs—these could be the subjects of many books! Following coral reef ecology is a chapter on the current status of corals in the Caribbean. This time of rapid human population growth, economic development, and global warming has produced a suite of very serious environmental challenges for the nations inhabiting the Caribbean region. Because of the importance of coral reefs for coastal protection, tourism, and fisheries, the survival of the corals that build the reefs is a critical part of these challenges. Finally, I have included a short section on

1

2 —o

Atlantic Ocean

3

4

Gulf of Mexico

7

5

6

11

12

8

9   10

13   15

Caribbean Sea

24

23

14

16

20

22   25

26

17   18

19

21

27

Pacific Ocean

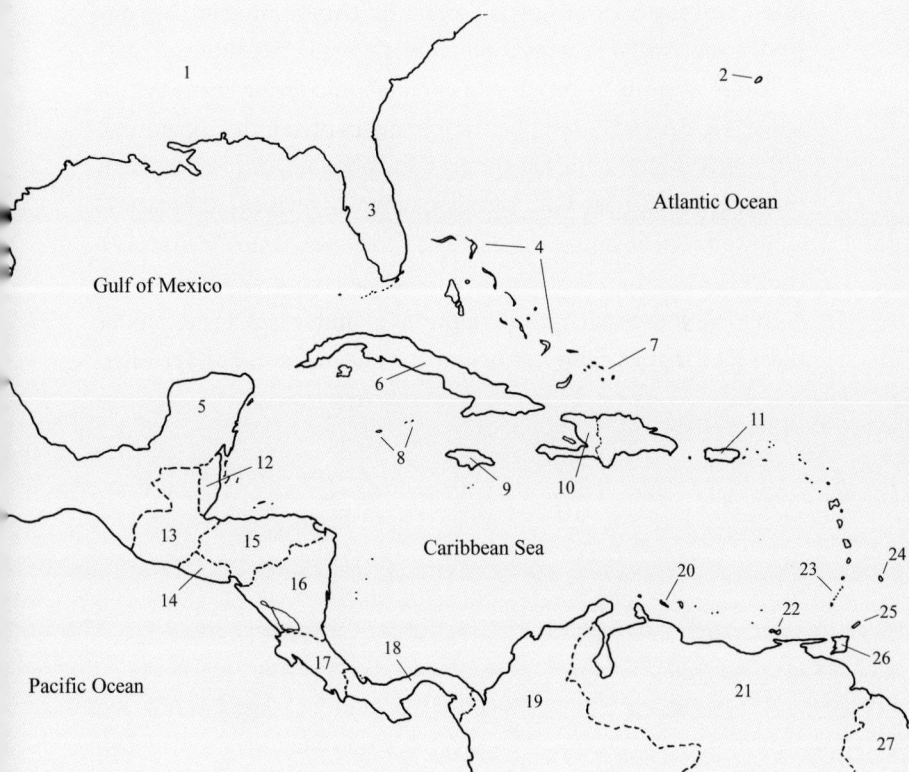

Map 1. The Wider Caribbean from roughly 8°N to 35°N and 58°W to 100°W. Numbers, roughly in rows left to right, relate to major countries and to islands mentioned in the text, as follows: 1 United States, 2 Bermuda, 3 Florida, 4 The Bahamas, 5 Mexico, 6 Cuba, 7 Turks and Caicos Islands, 8 Cayman Islands, 9 Jamaica, 10 Haiti, 11 Puerto Rico, 12 Belize, 13 Guatemala, 14 El Salvador, 15 Honduras, 16 Nicaragua, 17 Costa Rica, 18 Panama, 19 Colombia, 20 Curaçao, 21 Venezuela, 22 Margarita Island, 23 The Grenadines, 24 Barbados, 25 Tobago, 26 Trinidad, 27 Guyana.

the fossil history of corals, partly to satisfy the curiosity of those who have noticed the wealth of fossil corals in the coastal rocks of many Caribbean beaches, but also to point out that although corals have a very long history, coral reefs have not always been an important feature of tropical coastlines, suggesting that the continued survival of present coral reefs is not guaranteed.

Those wanting to get close to corals should ideally learn to dive with SCUBA (self-contained underwater breathing apparatus), but much can be achieved simply by snorkeling, or even through the bottom of a glass-bottomed boat, the windows of a submarine, or the glass of an aquarium (the last three should surely stimulate the desire to progress at least to a mask and snorkel). All you need is to look and watch, and perhaps take photographs with any of the wide range of cameras currently available for use underwater.

Corals of Florida and the Caribbean

# 1 | The Biology of Corals

## What Is a Coral?

The commonest application of the term coral is to the large, colonial, hard corals that have built tropical coral reefs. All hard corals have a rigid skeleton made of calcium carbonate (limestone), and most belong to a group called the Scleractinia (or stony corals). These, along with other hard corals, are classified within the Cnidaria, a major zoological group (phylum) that also contains sea anemones and jellyfish. The Cnidaria are so named because of the cnidae or stinging organelles they all possess. All Caribbean corals can sting, but with the exception of fire corals, most are unable to sting humans, because the delicate stinging threads shot out by their cnidae cannot penetrate human skin. Cnidaria also have other special characteristics, in particular, a simple body structure: a two-walled cylindrical sac with only one entrance, the mouth, surrounded by tentacles. This basic structure, in the case of corals and sea anemones, is referred to as a polyp. So-called reef-building corals contain zooxanthellae (microscopic single-celled algae) within their tissues, living as symbionts and assisting in the processes of nutrition and growth of the coral. This property, however, is not limited to corals, nor are zooxanthellae found in all corals.

Plate 1.1. Several species of Caribbean gorgonians flexing with wave surge, 10 m deep, Pedro Bank, south of Jamaica.

1

Non-scleractinian hard corals are represented on Caribbean reefs by a few species of a different group of cnidarians, the Hydrozoa. These differ from scleractinian corals (and from sea anemones) by having a much simpler body structure, but the chief difference noticed by swimmers between these and scleractinian corals is that the hydrozoans sting when you bump into them! They are the fire corals, very common and ecologically important on Caribbean reefs. Like other reef builders, fire corals also contain zooxanthellae.

Scleractinian corals that do not form typical coral reefs include those living in cooler and deeper water. Corals living in deep water are very diverse and include a range of forms, from simple, solitary "cup-corals" to extensively branched colonies. They lack zooxanthellae because of the lack of light in the deep sea, but some species form reefs in the deep, cold water. These deep reefs have only recently been discovered and their high biodiversity described, but they are already being threatened by deep-sea fishing. Other hard, calcified but non-scleractinian corals from cool or deep habitats include precious corals such as the Mediterranean red coral. Precious corals are distinctive in that their skeletal structure is dense and can take a high polish. In contrast, most scleractinian coral skeletons are porous and full of holes and cannot be polished to make high-quality jewelry.

The term coral, however, is also applied to various other cnidarians that do not have a rigid skeleton. These include the gorgonian corals (abundant on Caribbean reefs, plate 1.1), soft corals (common on Indo-Pacific reefs but rare in the Caribbean), and precious black corals (occurring on deeper parts of coral reefs). These have plantlike growth forms, are not rigidly calcified, and flex with underwater currents like miniature trees in a high wind.

Finally, the adjective coralline is applied to various organisms with hard calcium carbonate skeletons that are not corals by any definition but may live on typical coral reefs. Of these, crustose coralline algae are the most conspicuous; they grow as thin pink encrustations, but some form thicker coral-like lumps

Plate 1.2. Two species of Caribbean black coral: unbranched wire corals (*Stichopathes leutkeni*) in the foreground, with a finely branched species (*Antipathes caribbeana*) behind them.

or branches. They help to build reefs by cementing broken coral skeletons together, and by contributing sediments to fill spaces in the reef framework.

## Coral Structure

Most scleractinian corals living on coral reefs are colonial animals, meaning that each coral is made up of numerous more or less identical polyps that have a degree of independence but are directly connected to each other by living tissue, and are all

Plate 1.3. Pink coralline algae encrusting dead coral and barnacles (*above*), with yellow-brown blade fire coral *Millepora complanata* (*below*).

Plate 1.4. Contracted polyps of the great star coral, *Montastraea cavernosa*, by day; polyps are about 8 mm in diameter.

genetically identical. All polyps of a colony share resources and are integrated by a diffuse nervous network. The body of a coral therefore consists of many polyps covering a skeleton that is usually attached below to a hard surface such as a rock. The tissues overlie the skeleton and take its shape when the polyps are contracted (plate 1.4). In this state the living tissue forms a thin slimy film over the skeleton, and skeletal details can sometimes be seen through the tissues. To expand, the polyps swell with water, extend their tentacles, and rise above the skeleton, obscuring its features (plate 1.5).

The detailed structure of the skeleton is dictated by the polyps, each one sitting within and upon its own skeleton, called a corallite. Each corallite has more or less porous walls, with the space between adjacent corallites filled by porous skeleton or

Plate 1.5. Polyps of *M. cavernosa* extended at night.

coenosteum. Radiating from the central region of each corallite are numerous thin vertical plates called septa, usually in multiples of 6 (figure 1.1). The septa vary in size, the biggest (often 6) being called primary septa, the next biggest (6) secondaries, the next (12) tertiaries, and so on. The outer edges of the septa attach to the corallite walls while the inner edges may attach to a central structure called a columella. Septa can project beyond the corallite walls, either out of the top or sideways through the skeleton, with the parts outside the walls called costae and the whole a septo-costa. The free edges of the septa may be ornamented with granules, spines, or lobes; when these form near the columella they are called paliform lobes or, if arising from fused septa, pali. When the tissues of the polyps are contracted they are pulled into the corallites, folding between the upright septa. The depth to which the tissues go marks the bottom of the corallite; here there may be a thin plate separating the outer, living part of the coral skeleton from the old skeleton beneath.

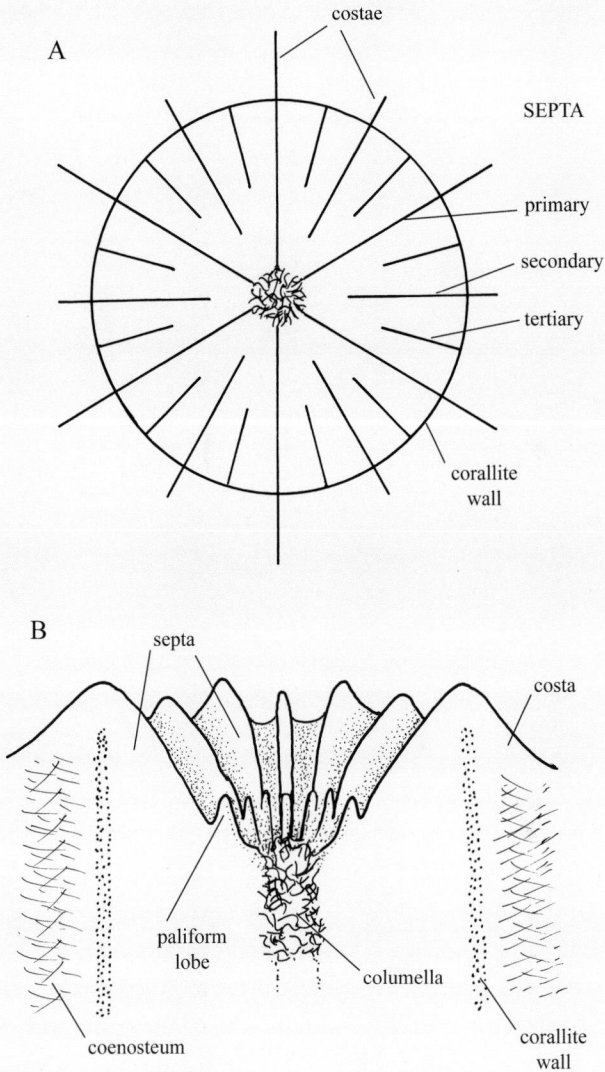

Figure 1.1. Skeletal features of a corallite. *A*: Diagrammatic plan view of a corallite showing septa, columella, corallite wall, and costae. *B*: Representation of a longitudinal section through a corallite showing septa with paliform lobes.

Plate 1.6. Skeleton of lobed star coral *Montastraea annularis*, showing corallites (about 2–3 mm in diameter) with primary and secondary septa and central columella.

The polyps themselves, as mentioned above, are anatomically simple, each consisting of a much folded sac sitting upon its corallite. The mouth is at the center of the outer end, surrounded by tentacles that are hollow, blind-ending tubular extensions of the wall of the sac. The inside of the sac is the digestive cavity, and this is joined to the digestive cavities of neighboring polyps by narrow channels connecting through the coenosteum. Digestion and absorption of food are performed by the cells lining the digestive cavity, with any waste passing out through the mouth. Polyps are extended by pumping up the sac, including the tentacles, with seawater taken in through the mouth. This gives the polyp and tentacles a hydrostatic skeleton that enables bending and stretching movements using muscles. The muscles, nerve network, and reproductive tissues lie between the two cell layers forming the walls of the sac; there is no circulatory system other than the fluids passing through the digestive cavities.

So far, the description of polyps and skeleton has referred to the most common arrangement, in which there are separate polyps and corallites. In many corals, however, there is less clear separation, and in some species, notably the brain corals, the polyps and corallites run into each other in elongated grooves lined on each side by septa (plate 1.7). In the living coral there are rows of mouths along the bottoms of the grooves and, when expanded, tentacles on either side (plate 1.10). These grooves branch and meander, and corals with this arrangement are described as meandroid.

Plate 1.7. Skeleton of maze coral, *Meandrina meandrites*, showing the meandroid arrangement without separate corallites.

# Growth

Colonies grow by secretion of new skeleton beneath the polyps and formation of new polyps. Secretion of the calcium carbonate skeleton is a complex process in which the ingredients are obtained from both the surrounding seawater and from internal metabolic processes. It proceeds faster in light, indicating some role played by the photosynthetic activities of the zooxanthellae (Gattuso et al. 1999; Furla et al. 2000). Skeletal growth and tissue growth proceed together, increasing the space for new polyps, and these arise either by division of existing polyps (referred to as intratentacular budding) or by differentiation from the tissue between polyps (extratentacular). The new polyps are therefore genetically identical to all others in the colony.

Corals are slow-growing animals. Some are *very* slow, with rates of growth of less than 1 cm per year. Even the fastest—the branching corals—only rarely exceed 15 cm per year. Variations in growth often occur on a regular, annual basis, resulting in formation of annual growth rings or bands. These can be used to estimate the age of a coral by cutting through the skeleton and counting the rings, as in a tree trunk. Rings are most clearly seen using X-rays on thin slices but, as shown in plate 1.8, can be roughly seen without expensive apparatus. Coral colonies several meters across may be hundreds of years old. In large, living colonies, it is possible to determine age without killing the coral by drilling out a core a few centimeters in diameter and counting the bands along the core. After drilling, the hole can easily be plugged and tissue may grow back over the plug. The annual bands are caused by changes in density of the skeleton, with denser skeleton (i.e., dark bands) representing seasonal times of slower growth, when calcium carbonate deposition continues but is not matched by skeletal extension. This can occur, for instance, during breeding seasons when the energy of the animal is applied to reproduction instead of growth (Mendes 2004). Care must be taken when interpreting the bands, however (hence the use of X-rays), as extra bands (slower growth) in a single year may be caused by irregular events such as coral bleaching, pollution, or other stresses.

## Growth Forms

The diverse shapes or growth forms of corals arise through different rates of skeletal growth in different parts of the colony. The four main growth forms—massive, branching, platelike, and encrusting—are broad descriptive terms for the overall colony shape or habit but overlap with each other and conceal much variation within each. The massive morphology is the simplest; in its pure form it is a part-sphere resulting from equal radial growth in all directions except downward. The closest approach to this highly symmetrical growth form is found in some brain corals. Other

Plate 1.8. On a rocky beach in Jamaica, view of a fossil coral eroded along the plane of growth (longitudinal section) and showing about 30 annual growth bands.

versions of massive include low domes and irregular lumps; these need not be large to qualify as massive: the criterion is shape rather than size. It would not seem, therefore, that massive would overlap in any way with branching, yet in fact it does in some large colonies formed of thick columns, such as the lobed star coral *Montastraea annularis* where small colonies are clearly massive (one short column or lump) whereas large colonies, when examined in detail, may turn out to consist of several to many thick branched columns with living tissue only at the tops—giving the appearance of a single huge, irregular lump. Pillar corals and encrusting corals with small irregular lumps also challenge unambiguous categorization into one of the four main morphologies, and it is often the case that a single colony may display more than one morphology.

Branching morphologies arise as a result of growth being fastest at the tips of branches and inhibited further back, so branches reach a characteristic thickness for the species in question. Other aspects of branching morphology, such as the places where secondary branches occur and the directions in which branches grow, are also controlled through differences in growth rates. Many different branching forms occur, with names that link appearance to more familiar growths such as thicket, tree, or staghorn. Platelike growth forms are flat sheets that may be aligned in any plane from vertical to horizontal. As in branching colonies, platelike growth is achieved by faster growth outward at the edge than growth in thickness further back. Mixtures of massive and platelike growth may be found in some species when a massive colony develops platelike skirts around the lower, side-facing edges, possibly in response to shading. Encrusting growth forms may be typical of a species, or simply a form adopted in response to the environment. Thus some colonies send up plates or branches from an encrusting base, and some species that normally form massive colonies may form crusts instead of domes in dim light. Similarly, platelike colonies appear to encrust when they are closely applied to the underlying rock.

Variations in growth form, both within and between species, illustrate the plasticity of colony growth and reveal subtle and intimate connections with the environment.

## Nutrition

Nutrition is achieved in at least two ways; each method varies in relative importance depending on the species of coral and the environment. The simplest major method is the capture of small animals by using the tentacles and stinging cells. Suitable small

Plate 1.9. Maze coral, *Meandrina meandrites*, showing polyps contracted by day; compare with plates 1.7 and 1.10.

Plate 1.10. *M. meandrites* with polyps and tentacles extended at night to feed; magnification × 2.

animal food is mostly to be found in the plankton and consists of adult and larval crustaceans, worms, mollusks, etc., each a few millimeters or less in length. These bump into the extended tentacles, stimulating the stinging cells to fire stinging or sticky threads that impale or entangle the prey, immobilizing it and allowing the coral to stuff it into the adjacent mouth. Many corals extend their tentacles only at night since plankton rises toward the water surface at night and is therefore commonest on the reef during darkness. Plankton sinks into deeper water during the day, probably because of the high risk of visual predation by small day-active fish. A diver on the reef at night can easily feed the corals by focusing a flashlight beam on a suitable colony, attracting a swarm of plankton to the waiting tentacles.

The second major method of nutrition is for the coral to absorb and use products of photosynthesis released by the symbiotic zooxanthellae. This represents a central aspect of the symbiosis between the coral animal and its contained plants. The zooxanthellae gain protection, a place in the sun, and fertilizer in the form of the excretory products of the coral's predatory activities, while the coral benefits by using a proportion of the photosynthetic output of the zooxanthellae for its nutrition. It is also possible that the skeletal growth of the coral is aided by the zooxanthellae, since during photosynthesis they remove carbon dioxide, a byproduct of the secretion of calcium carbonate.

Other possible supplementary feeding methods of corals include the capture of detritus (dead organic matter) and phytoplankton (tiny planktonic plants), and the absorption of dissolved organic matter (DOM) directly from the seawater. Although experimental evidence confirms that corals can gain some nutrition from these feeding methods, their contribution to the total food intake of most corals is rather small since these food sources are either scarce on typical coral reefs (phytoplankton and DOM) or not very nutritious (detritus).

In the case of the two main methods of nutrition, experimental and observational evidence suggests that both may be necessary for maximum growth and reproduction, but that in their natural habitat different species experience different proportions of each. Most coral species, for instance, occur over a wide depth range, and the intensity of light, and therefore of photosynthesis by the zooxanthellae, decreases with increasing depth. Thus less photosynthetic nutrition is available to deeper living corals, which therefore need either to catch more planktonic food or to reduce their metabolic rates. It has been shown that deeper-living corals do have lower metabolic rates (and slower growth rates) than those in shallower water. Reducing light intensity is clearly the main factor limiting reef-building corals to relatively shallow water (usually less than about 50 m deep).

## Reproduction

Corals can reproduce both sexually and asexually. Asexual reproduction commonly occurs when branching colonies are broken by waves and some of the scattered branches regrow to form new colonies (plates 1.11 and 1.12). This process is more successful for larger broken pieces than for smaller fragments, which tend to roll about, killing all the tissues. Larger pieces may get wedged among others, giving sufficient stability for surviving tissues to grow and secrete more skeleton. Encrusting skeletal growth binds pieces together, increasing the stability of the new colony. Other less accidental methods of asexual reproduction include production of small balls of tissue that are shed from colonies. These may attach nearby to form new colonies.

Asexual reproduction produces new colonies that are genetically identical to the parent colony. In contrast, sexual reproduction produces nonidentical offspring through mixing of the genes of the parents. Parental genes are shuffled and recombined during the type of cell division (meiosis) that produces the eggs and sperm, each of which carries a single set of parental genes. When an egg is fertilized by a sperm to produce a new individual, one set of genes is combined with a different set from the other parent. This mixing of genes is seen as advantageous to the species since it results in increased genetic diversity, making it more likely that the species will survive environmental changes since some of the genetic variants may have traits that are favorable in the changed circumstances. The enhanced survival of favorable genetic variants is, of course, the basic mechanism of evolution.

Sexual reproduction in corals takes many forms (Richmond 1997). Some species are hermaphrodites whereas others have separate sexes (gonochoric). Common examples of hermaphroditic

Plate 1.11. Staghorn coral, *Acropora cervicornis*, showing survival of tissue on branches broken in a hurricane.

Plate 1.12. On a rocky platform, fragment of elkhorn coral, *Acropora palmata*, showing new and reoriented growth.

species include *Acropora cervicornis*, *Diploria strigosa*, and *Montastraea annularis*. Gonochoric species include *Dendrogyra cylindrus*, *Montastraea cavernosa*, and *Siderastrea siderea*. All of these are spawning species, meaning that they release bundles or clouds of sperm and eggs (often together) into the water where they float to the surface and drift with the currents (Richmond 1997; plates 1.13 and 1.14). Fertilization takes place in the water, and the fertilized eggs develop into tiny balls of cells called planula larvae or planulae. These are able to swim slowly using cilia (tiny subcellular hairs) and have sensory capabilities that enable them in due course to find suitable sites to settle and develop into new coral colonies.

Spawning is highly synchronized: often restricted to just a few nights each year. In fact, colonies of a single species do not synchronize spawning just with each other, but also with other species, so a multispecies slick of eggs and sperm may accumulate at the surface during a mass-spawning event. These are less spectacular in the Caribbean than in the Indo-Pacific, but nevertheless they occur on Caribbean reefs, generally over a few nights in the late summer. Timing appears to be governed by increasing water temperature over the summer, which controls the development of the gametes, and the light of a particular phase of the moon to trigger spawning. Synchronization within a species is important to ensure successful cross-fertilization, but mass spawning of several different species probably also has the advantage of

Plate 1.13. A cloud of gamete bundles above a spawning coral, *Montastraea* sp. (Photo: J. C. Ogden.)

Plate 1.14. Spawning of *Montastraea* sp., showing gamete bundles being released from the mouths of polyps. (Photo: J. C. Ogden.)

overwhelming egg-predators—many eggs will be eaten, mostly by fishes, but many more will escape the satiated diners.

In contrast to the spawning species, there are also coral species that are brooders; these release sperm into the water but retain the eggs inside the body. The eggs are fertilized by sperm taken in from the surrounding water, and the fertilized eggs develop into planulae before being released from the parent. Once released, they are able to settle and develop into new colonies either immediately or after a time drifting with the plankton (figure 1.2). Common examples include *Agaricia agaricites* and *Porites astreoides*, both hermaphroditic; and there are also gonochoric brooders, e.g., *Siderastrea radians*. Since the eggs are retained to wait for the sperm, there is less need for brooders to synchronize their breeding as precisely as spawners do. And since fertilized eggs are also retained for a time, a lower rate of predation is likely. However,

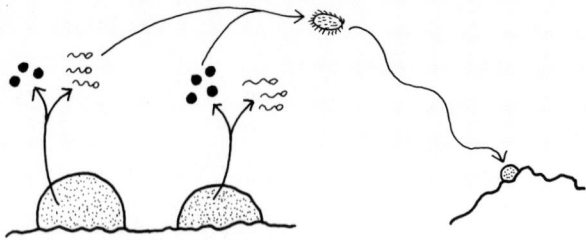

a - hermaphroditic spawning

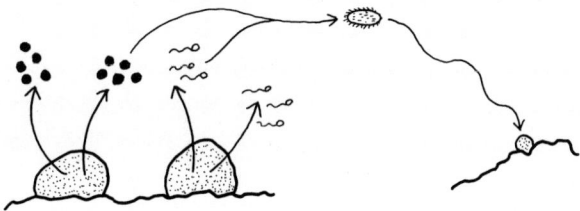

b - gonochoristic spawning

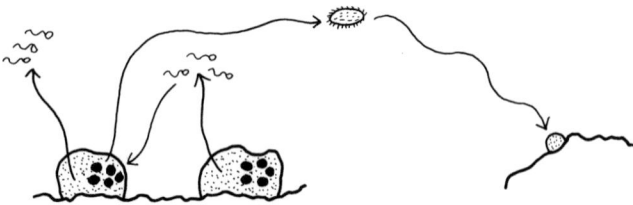

c - hermaphroditic brooding

Figure 1.2. Life cycles of spawners and brooders. Fertilization of eggs (represented as black spots) by sperm (open circles with tails) lead eventually to planula larvae that settle and develop into juvenile corals.

there appears to be no overwhelming advantage to either brooding or spawning; although spawning is commonest, both reproductive strategies occur in reef coral communities with abundant species represented in each (Richmond and Hunter 1990; Richmond 1997).

Recruitment is the final stage of reproduction and consists of the establishment of a small colony on the reef. It starts with the settlement of the planula. This is a fairly complex process since the planula needs to find a suitable spot, judged by such factors as depth, light, water movement, and the texture and chemical makeup of the surface. Some corals prefer to settle on particular species of encrusting coralline algae, while others are less specific. Commonly the surface should be rough for secure attachment, with some water movement, and although sufficient light for photosynthesis is necessary, there is a preference for cryptic locations such as crevices or downward-facing surfaces. Once attached, the planula must change into a polyp, start to secrete a skeleton, and, if it does not already have them, acquire zooxanthellae from the surrounding water. Several things can go wrong while this metamorphosis is occurring: a grazing animal may pass by and eat the baby coral, a piece of debris may wash past and dislodge it, it may be buried by a layer of sediment or smothered by algae, or it may run short of energy and simply die. Only when it has avoided such dangers and has grown big enough to defend itself and continue growing (perhaps at 1 or 2 cm across) can it be counted as a recruit. Unsurprisingly, rather few individuals survive their time in the plankton and the hazards of settlement and recruitment, so the numbers of recruits counted on Caribbean coral reefs are usually rather low (a mean of about 3 per square meter, Marks 2007).

## Behavior

Apart from predatory feeding, corals do not appear to do much, and even the feeding takes place mostly at night; however, on flourishing reefs the corals are constantly engaged in a slow-motion battle with their neighbors (Lang and Chornesky 1990). This is a battle for space and light, and the enemies are all other sessile (fixed to the substrate) living things, including other corals, sponges, and algae. During normal growth, corals and other sessile organisms inevitably encounter each other and may overgrow or overshadow each other. Clearly this is liable to restrict growth and can lead to partial or complete mortality, so corals need defensive abilities.

Plate 1.15. Staghorn coral, *Acropora cervicornis*, overtops lobed star coral, *Montastraea annularis*. Photographed in Jamaica in 1976.

Fast growth is one strategy to overgrow neighbors, and this is used by branching corals to grow above and shade less rapidly growing massive corals; however, since branching corals are more easily broken by storms, the victory so gained is often only temporary. Another strategy is to be more aggressive than the neighbors. This involves stinging them and attempting to kill them so their space and light can be taken. Aggressive encounters in corals generally take place at night when the polyps are expanded, and special long tentacles (sweeper tentacles) may be used to reach out and sting an encroaching neighbor. More often, however, white mesenteric filaments with powerful stinging cells are extruded from the body cavity, and these stick to the tissues of the neighbor, killing it if possible. As in all other respects, coral species vary in their competitive abilities; while some species succeed by stinging, others succeed by faster growth.

The other side of aggression is of course defense, and there is much to defend against besides neighboring sessile organisms. Settling larvae are a potential problem, and so are predatory animals; potential burial by settling sediment is another danger. Corals defend themselves against other organisms by using their stinging cells, their skeleton, defensive chemicals, and mucus (slime). The whole outer surface of a coral is supplied with mucus-producing cells and is also ciliated, allowing the mucus to be moved. It is likely that some potential predators are deterred by such crunchy, slimy, stinging, and unpleasant-tasting food. Settling larvae may also be stung, poisoned, or wrapped in mucus and possibly eaten. Settling sediment is wrapped in mucus and moved by cilia to the edge of the colony to be shed. In some environments, strong defensive capabilities or heightened ability to shed sediment may give a competitive advantage to particular species, outweighing mere stinging ability or growth rate. Variability such as this is responsible for the high species richness of coral reefs because it ensures that no one species can take over the whole habitat.

Plate 1.16. Boulder brain coral, *Colpophyllia natans*, overgrows symmetrical brain coral, *Diploria strigosa*, and its Christmas tree worms.

# 2 Species Descriptions

The species described in this book include almost all those that occur on coral reefs in the wider Caribbean and likely to be encountered by divers or snorkelers. All but a few are reef-building corals that contain zooxanthellae in their tissues. In putting together this selection of corals I have used two main guidelines: to exclude rare or inconspicuous corals, at least from full description, and to take a conservative position on species.

This latter position requires explanation. People who describe and name animals and plants (taxonomists) have to decide whether slightly different varieties of a species are actually just varieties or are real and distinct species in their own right. In zoology, the distinctions between populations of animals that justify separating them into different species include the existence of several distinct, nonoverlapping, morphological or biochemical differences between the various populations. Such differences strongly suggest that the populations do not interbreed in their natural habitat since, if they did, intermediates would frequently be found. It is assumed that they have developed their differences by evolving along separate paths. If the differences overlap, that is, if there *are* intermediates, this suggests that the relevant populations are not reproductively isolated and thus represent varieties of a single but variable species (sometimes called a species complex) in which some hybridization between varieties occurs. Taxonomists differ in their approaches to this problem: some view small differences as important and sufficient to separate species, whereas others tend to group together animals with what they regard as only small differences—the two approaches are often referred to as splitting and lumping, respectively. The modern way of deciding is to use molecular biology to investigate genetic differences between populations. Several problematic species complexes have been subjected to this analysis, but questionable species distinctions among common corals remain unresolved or unstudied at the genetic level. Even genetic distinctions require interpretation, and in any case evolution predicts the existence of intermediates during speciation. So the problem remains for those whose job it is to name things. My own position is that I prefer to lump rather than to split, especially where molecular genetic evidence supports such a view. In the following descriptions I point out where difficulties currently exist, but the pace of research in this area is such that many of these difficulties may soon be resolved and perhaps others revealed.

Like all animals and plants, the scientific names of coral species comprise two names—genus and species, usually written in italics with the first letter of the genus capitalized (e.g., *Montastraea annularis*). Species believed to be closely related are placed in the same genus. Higher-order groupings of species, ideally reflecting evolutionary relationships, include families, orders, and classes; thus, genera (plural of genus) believed to be related are placed in the same family, and I have arranged the species descriptions in families to help show the characters common to each. In fact, however, as with species, so with genera and families, experts disagree! Taxonomists may move species from one genus to another, or from one family to another, on the basis of new information. Classical coral taxonomy rests mainly on characters of the skeleton, since this is often the only part that remains once specimens have found their way to a museum (and in the case of fossil corals, is the only part available anyway!). Examination of live animals, soft parts, cells, molecules, and particularly molecular genetics provides a much greater source of information about possible differences and similarities, so a constant stream of new information is emerging, altering our present understanding. Here, the order of presentation of families and the placement of species within particular families broadly follow Veron's (2000) three-volume *Corals of the World* except where new information suggests more probable arrangements; however, the considerable amount of new data from molecular genetics suggests that many, perhaps most, existing coral families, and some coral genera, are polyphyletic (containing two or more different evolutionary lines) and thus that traditional coral taxonomy requires a serious makeover. Of particular interest here is that many Atlantic (Caribbean) coral species are not as closely related to Indo-Pacific species as might be supposed from their present inclusion in the same genera or families (Fukami et al. 2004b). At present, no agreed new arrangement of species within monophyletic genera (species within each genus descended from a single ancestral species) or of genera within monophyletic families is available to replace the existing traditional taxonomy (Fukami et al. 2008; Kitahara et al. 2010).

Until a few years ago, most corals did not have common English names, and even those names that did exist were often used for more than one species (e.g., brain coral). This confusing situation changed in 1993 with the publication of Paul Humann's *Reef Coral Identification: Florida, Caribbean, Bahamas.* Humann chose commonly used but separate descriptive names for each species, and if none existed, he invented one. I have followed the lead of Humann and DeLoach (2002) in the second edition of the book and used their common names for the various species.

We are relatively lucky in the Caribbean when it comes to identifying species of corals since there are comparatively few of them, and even fewer common species. Depending on how you divide species, between 60 and 80 species have been described from the Caribbean region. In contrast, in the Indo-Pacific region, more than 10 times this number have been described. This difference is due partly to the much greater size of the region—more geographical space generally leads to the evolution of more species— but also to the geological history of the Caribbean, especially the environmental changes consequent on the separation of the region from the Pacific. This history is described in chapter 5 on fossil Caribbean corals.

Together with the species descriptions below, I have included short sections on families and genera where these seem appropriate, with the descriptions themselves organized within families. All descriptions include information on the morphology, size, color, and habitat of the species in question. These data are based on a range of taxonomic publications (Wells 1973; Cairns 1982; Fenner 1993, 1999; Veron 2000) supplemented by personal observations. Abundance is indicated by a rough descriptive scale from rare through uncommon, occasional, frequent, and common to abundant and dominant. It should always be recognized, however, that species may vary in their geographic ranges, so a coral that is rare or absent in some places may be common in others.

# Class Hydrozoa

## ◼ Family Milleporidae, fire corals

Fire corals can easily be distinguished from all scleractinian corals by the absence of both corallite skeletal structures and anemone-like polyps. The surface of fire corals appears smooth, and even under low magnification all that can be seen are many small holes or pores. In life, fine tentacles carrying powerful stinging cells project from these holes, one from each. The tentacles can be revealed by close-up photography as a forest of hairs projecting all over the coral's surface. Interspersed among the numerous longer tentacles are fewer shorter ones bearing the mouths of the colony. Anatomically each of these tentacles is a single polyp—specialized either as a catching tentacle (without a mouth) or as a feeding polyp (with a mouth). The arrangement is often such that the mouth-bearing polyps are surrounded by a ring of tentacle polyps. The colors of fire corals range from yellowish through rich

Plate 2.1. Branching fire coral, *Millepora alcicornis*, showing extended tentacles.

Plate 2.2. Closer view of *M. alcicornis'* extended tentacles.

orange-brown to pinkish, with the growing tips or edges being paler or white.

While the Caribbean fire coral species described below appear different from each other, it is often possible to find intermediate growth forms that are difficult to assign to any particular species. There is thus some doubt over the validity of some Caribbean fire coral species. Molecular biological research so far undertaken supports this doubt, suggesting that there are actually fewer unambiguous species than were thought to exist (Ruiz Ramos 2009).

### *Millepora alcicornis*, branching fire coral (plates 2.1–2.5)

Both this and blade fire coral, described below, occur very commonly on Caribbean reefs. Both live in shallow water about 1–15 m deep. Branching fire coral is most common at the deeper end of this range, whereas blade fire coral generally occurs in shallower water. As the name suggests, the chief feature by which

Plate 2.3. *M. alcicornis* colony forming an irregular bush.

Plate 2.4. *M. alcicornis* encrusting a sea fan, with unencrusted gorgonians in the background.

*M. alcicornis* can be distinguished from other fire corals is its branched structure—sometimes bushlike (plate 2.3) or shaped as a fan, extending up to about 10 cm from its base. Fans are usually oriented at right angles to the normal direction of water movement—waves or currents (plate 2.5)—giving the coral a greater chance of catching food streaming past in the water.

The individual branches of branching fire coral are fairly thin (2–8 mm diameter), but the basal sheet from which the branches spring can cover areas of 0.5 m² or more by encrusting the surface on which the colony is growing, sending up further branched fans or clumps at frequent intervals. One of the most impressive encrusting tricks of branching fire coral is that of covering the whole surface of a dead sea fan; indeed, the fire coral may progressively kill the sea fan as it encrusts. The growing coral faithfully follows the contours of the fine branches of the sea fan so that it eventually looks quite like a normal sea fan, but a bit stiff and the wrong color (plate 2.4).

In deeper water, *M. alcicornis* branches become thicker and fewer, taking on an encrusting lumpy morphology that makes it increasingly difficult to distinguish from other species.

### *Millepora complanata*, blade fire coral (plates 2.5–2.6)

This species is distinguished from the last by growing as thin, flat plates; at their edges the blades can be just 1 mm thick. The plates or blades rise up to about 10 cm from an encrusting base and, like those of the previous species, are generally parallel to each other

Plate 2.5. Fan-shaped blade fire coral, *Millepora complanata* (*left*), and fan-shaped branched fire coral, *M. alcicornis* (*right*), growing among green calcareous algae.

Plate 2.6. Blades of *M. complanata*, with finger coral, *Porites porites*, on the left.

and oriented at right angles to the direction of water movement (plates 2.5, 2.6). Blade fire coral flourishes in shallow water, often in areas exposed to moderate wave action so that surf breaks over the living coral. Heavy surf of course breaks the longer blades, but the coral heals quickly, and the broken pieces can re-cement themselves to suitable rocky substrate nearby and start new colonies. Snorkelers in shallow water near reef crests need to be especially watchful to avoid being washed by waves into painful contact with this species, which can be abundant on the outer edges of reefs.

Some authorities have described fire corals with vertically corrugated or striated blades as a separate species, *Millepora striata*. Recent morphological studies and molecular evidence, however, suggest that not only is this form not distinct from *M. complanata*,

Plate 2.7. Box fire coral, *Millepora squarrosa*, with sponges and algae growing between lobes, The Grenadines, depth of about 4 m.

but that both are very close to *M. alcicornis* (Ruiz Ramos 2009). Probably all three form a single, morphologically variable species complex.

*Millepora squarrosa*, box fire coral (plate 2.7)

Box fire coral grows in shallow water in somewhat similar environments to those inhabited by blade fire coral; however, its growth form is different in that although it consists of blades arising from an encrusting base, the blades are much shorter, thicker at their edges, and joined to each other in a reticulated pattern to form many small boxes, each up to a few centimeters in diameter. The species has a restricted distribution in the Caribbean, being found predominantly toward the east, especially in the Lesser Antilles. Molecular evidence suggests that this species is distinct from the *M. alcicornis* and *complanata* species complex (Ruiz Ramos 2009).

## ◼ Family Stylasteridae, lace corals

Unlike fire corals, the lace corals do have structures that look like corallites, but because they are hydrozoan corals, these structures are quite different from true scleractinian corallites. One species is fairly common on Caribbean reefs.

*Stylaster roseus*, rose lace coral (plate 2.8)

Rose lace corals generally appear as small fan-shaped colonies 3–6 cm in diameter, hidden away in cracks and crevices of the reef. As the name suggests, the colonies are composed of many fine branches, giving a lacy appearance; the color ranges from purple

Plate 2.8. Rose lace coral, *Stylaster roseus*, in a crevice.

through pink to white. As with the fire corals, the polyps of *S. roseus* are either tentacle polyps or mouth-bearing, feeding polyps. These polyps are strictly arranged in tiny corallite-like cups on the surface of the branches; in each, a circlet of tentacle polyps surrounds a central feeding polyp. The appearance of the tentacles in close-up photography is similar to that of fire corals, but instead of projecting evenly from the surface, the tentacles radiate from the cups.

These corals lack zooxanthellae and are probably restricted by competition with algae and scleractinian corals to unlighted parts of the reef. In shallow water (1–10 m) they occur in crevices or within the reef framework, where the wash of waves drives water past them. In deep or murky water, however, and especially where there are water currents, they can be found projecting from the open rock surface with the fans oriented perpendicular to the current direction.

## Class Anthozoa, Order Scleractinia

### ■ Family Acroporidae

This family contains four genera, only one of which, *Acropora*, is found in the Caribbean, where it is represented by two (possibly three) species. In contrast, in the Indo-Pacific region the genus is extremely diverse with some 150 described species spanning a wide range of colony forms. Because of this Indo-Pacific diversity, *Acropora* is the most species-rich of all coral genera. Until recently, although only represented by two species on Caribbean reefs, *Acropora* species were the commonest shallow-water Caribbean corals; however, a drop of more than 90% in abundance of *Acropora* occurred throughout the Caribbean during the 1980s. This huge drop is thought to have been caused mainly by a coral disease called white band disease, discussed in chapter 4. *Acropora* species are branching corals, characterized by small, protruding, relatively simple corallites and porous skeletons. They are fast growing, often exceeding 10 cm linear growth of branches per year.

*Acropora cervicornis*, staghorn coral (plates 2.9–2.10)

As its scientific and common names imply, the branches of this coral resemble stag's antlers. Branches are generally 1–2 cm in diameter and fork repeatedly to form thickets extending up to 1 m above the substrate and several meters wide. The live color is rich brown and the branches are rough in texture, being covered by the small, protruding, almost tubelike corallites. Each branch bears an apical corallite at the growing tip, larger than the radial corallites budded from it and pale or white in color, an indication

Plate 2.9. Small thicket of staghorn coral, *Acropora cervicornis*, at depth of about 8 m, Jamaica.

Plate 2.10. Closer view of *A. cervicornis* branches, showing protruding corallites.

of its rapid growth. The network of branches in a staghorn thicket provides excellent cover for small fish that retreat within when predatory fish come by.

The habitat of this coral is at about 5–10 m depth on the seaward side of reefs, where its recent abundance led to the naming of this part of the reef the *cervicornis* zone, and also in shallower, more sheltered places behind the reef crest or in the lee of islands or headlands. On present reefs, staghorn coral is rarely common but can usually be found in small scattered patches. The loose, branching growth form is easily broken by waves, but this breakage can lead to an increase in the number of colonies, since the larger fragments may survive and grow in the new locations where the waves transport them. Indeed, fragmentation and regeneration is thought to be an important reproductive mechanism in this species. Unlike those of other corals, the colonies are usually not firmly attached to underlying rock and the thickets can grow, supported by dead basal portions, on sand or rubble—the latter often largely composed of dead and broken staghorn coral.

*Acropora palmata,* elkhorn coral (plates 2.11–2.13)

This coral forms robust treelike colonies up to 2 m high and 30 cm or more in thickness at the base, with thick, more or less flattened branches angled and facing upward and extending 1–2 m to the sides; the colony surface is rough with numerous protruding corallites (plate 2.13). Elkhorn coral is rich orange brown in color, with the relatively large corallites at the growing branch tips being pale or white. The branches vary in shape according to the degree of exposure to wave action, being flattened horizontally into broad plates in sheltered conditions (plate 2.12), and more

Plate 2.11. Abundant elkhorn coral *Acropora palmata* in the *palmata* zone of Curaçao.

rounded in cross-section like tree branches in exposed conditions. The branches tend to grow toward rather than across the direction of approaching waves. Colonies of this species used to cluster thickly to form a clearly defined zone—the *palmata* zone—in about 1–5 m depth on most Caribbean reefs facing oncoming waves; sadly however, although these zones still exist in a few places (e.g., Curaçao), many elsewhere are now composed largely of dead colonies still standing in life position but overgrown with algae, fire coral, and colonial zoanthids. As with the previous species, where *A. palmata* used to be abundant, live colonies can still occasionally be found.

Elkhorn coral can tolerate quite heavy wave action as it often lives very close to the water surface—snorkelers can barely swim over the *palmata* zone without risking serious scrapes. However, storm waves can break and topple the colonies, sometimes turning them upside-down. Because of their exposed habitat, elkhorn corals need firm attachment to rock for long-term survival, but large live pieces or toppled colonies, if heavy enough or wedged

Plate 2.12. Colony of *A. palmata* with flattened, platelike branches.

Plate 2.13. Closer view of *A. palmata*, showing protruding corallites.

among others, can grow and re-cement themselves to their rocky surroundings. New upward growth is reoriented with branches spreading out and angled toward the surface, and after a few years the regenerating coral will resemble a normal colony (plate 1.12).

A possible third species of *Acropora* is a form referred to as fused staghorn coral (*A. prolifera*), which most authorities think is a hybrid between staghorn and elkhorn corals. It is rather rare and looks like a cross between the two: either forming many branches in platelike fans, or plates made up of many partly fused branches. The existence of such a hybrid does not mean that elkhorn and staghorn corals are not themselves true species, since they are clearly very different, and the rarity of the possible hybrid suggests that such hybridization is an unusual event.

### ■ Family Astrocoeniidae

The corallites of this small family have regularly arranged, solid septa and stylelike columellae. It contains only one Caribbean genus, *Stephanocoenia*, which, in turn, contains only one species.

*Stephanocoenia intersepta* (syn. *S. michelinii),* blushing star coral (plates 2.14–2.16)

Colonies of *S. intersepta* grow as mounds or small coral heads fixed firmly to the underlying rock, 30 cm or more across and

Plate 2.14. Two colonies of blushing star coral, *Stephanocoenia intersepta*, with polyps contracted; the orange spots in the colony on the left are worm tentacles.

pale beige to brown in color. This coral derives its name from the fact that the numerous small brown polyps are often expanded by day, but if disturbed they retract, causing the coral to "blush" paler (compare plates 2.14 and 2.15). It is not usually found in shallow water, but below about 10 m it occurs quite frequently on most reefs and can be common in turbid reef environments. The small corallites (2–3 mm diameter) hardly protrude from the surface; they contain 12 primary septa, and 12 shorter secondaries. The primary septa have distinct paliform lobes that together form a circlet around the columella. Tiny developing corallites can often

Plate 2.15. *S. intersepta* with polyps extended (*left*); whitestar sheet coral, *Agaricia lamarcki* (*right*).

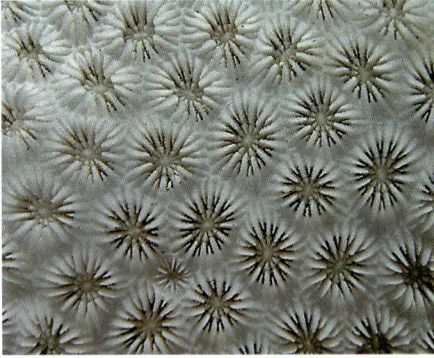

Plate 2.16. Skeleton of *S. intersepta*, showing corallites with both primary and secondary septa and paliform lobes.

be seen between the standard corallites, showing that this species grows by extratentacular budding.

This coral can be confused with small colonies of lobed star coral, *Montastraea annularis*, but has smaller, less protruding corallites and is generally paler in color.

## ■ Family Pocilloporidae

This family has several genera in the Indo-Pacific, but only one genus in the Caribbean—*Madracis*—moved from the Astrocoeniidae on the basis of molecular and microstructural evidence.

*Madracis mirabilis*, yellow pencil coral (plates 2.17–2.18)

Below about 10 m, or in sheltered shallower locations, yellow pencil coral can be very common and form large colonies covering several square meters. Its growth takes the form of numerous narrow branches, each about the thickness of a pencil and yellowish in color. Young colonies form a ball or dome of branches radiating from the center; further growth leads either to lateral extension to form a carpet up to 15 cm thick, or to breakage of branches at the edge, exposing the dead inner branch bases. Broken branches relocated by water movement or gravity may regenerate to form new colonies. The thin branches and fragile dead bases make this coral very susceptible to storm damage, which is no doubt the reason it is generally not found in shallow water.

Polyps are extended by day, filling the spaces between the clustered branches. Unusually, but like those of the next species, the small corallites (1–2 mm in diameter) have only 10 septa, all of which are fused to the columella.

Plate 2.17. Colony of yellow pencil coral, *Madracis mirabilis*, with sponges in the foreground.

Plate 2.18. Close-up of *M. mirabilis*, showing extended polyps.

## *Madracis decactis*, ten-ray star coral (plates 2.19–2.20)

*M. decactis*, named for its 10 septa, forms fairly small colonies, rarely extending beyond 10–15 cm wide and 8 cm high and composed of rounded lobes or short branches, 1–3 cm in diameter. Colonies generally appear greenish in life, but artificial light may reveal reddish tones between the paler polyps. The corallites are quite small (about 2 mm in diameter), with thick and obvious septa, but these may be obscured by the polyps, which are often extended by day. The species is quite common on reefs below about 5 m and extending down to 20 m or more; however, it tends to grow on vertical edges and in odd corners, so is easily overlooked.

Four other described species of *Madracis* may be encountered on reefs in the Caribbean. Colonies of *M. formosa* (eight-ray finger coral) are generally larger than those of *M. decactis*, grow in deeper water, and are much more clearly branched. The colonies are usually greenish, whereas the polyps have pale green or yellow centers. In addition, the corallites have only 8 septa. *M. carmabi* (ten-ray finger coral, plate 2.21) appears almost identical to *M. formosa* and also lives in deeper water, but is more brown than green

Plate 2.19. Ten-ray star coral, *Madracis decactis*, with polyps extended, surrounded by green algae.

and has 10 septa instead of 8. It may be a hybrid between *M. formosa* and *M. decactis* (Vermeij et al. 2003). In comparison with *M. decactis*, both *M. formosa* and *M. carmabi* are occasional or rare. *M. pharensis* (plate 2.22) is an encrusting form living in habitats similar to those of *M. decactis*. Like the latter, it often goes unnoticed but can be quite frequent, especially in the southern and eastern Caribbean. Its color is variable green to brown or purple and may be pale; in caves and crevices and in deeper water it may lack zooxanthellae. Like *M. decactis* it has 10 primary septa per corallite but also has 10 small secondary septa. Molecular studies have suggested that *M. decactis*, *M. formosa*, and *M. pharensis* are very closely related. They should perhaps be regarded as a species

Plate 2.20. Closer view of *M. decactis* with polyps retracted; red calcareous algae encrusts the bases of the coral.

Plate 2.21. Ten-ray finger coral, *Madracis carmabi*, at depth of about 30 m, off Discovery Bay, Jamaica. The colony is flanked by red encrusting calcareous algae.

complex in which the three morphologies, encrusting—slightly branched—more branched, form an overlapping series within which hybridization prevents reproductive isolation (Diekmann et al. 2001). If *M. carmabi* is also a hybrid, it should be included in this species complex. A final species, *M. senaria*, is rarely seen on account of its habitat in caves and crevices, often deep on the reef. It is an encrusting form like *M. pharensis* but has only 6 primary septa in its corallites.

Plate 2.22. Encrusting star coral, *Madracis pharensis*, with polyps extended.

## ■ Family Oculinidae

This family can be recognized by the tubular, well-defined corallites linked by solid, fairly smooth skeleton. The only genus represented in the Caribbean is *Oculina* (ivory corals), which includes several generally uncommon species that form small branching colonies. These species differ from each other mainly in the details of branching pattern, thickness of branches, and density and shape of corallites. Molecular genetic analysis would therefore be interesting to investigate whether these are local species or simply morphologies adapted to different environments. Only one species is described here.

*Oculina diffusa*, diffuse ivory bush coral (plates 2.23–2.24)

These delicate little corals form dense clumps of branches, up to about 15 cm in diameter. Each branch bears several or many widely spaced corallites, in roughly longitudinal rows or scattered over the surface. The corallites protrude slightly and are angled

Plate 2.23. Diffuse ivory bush coral, *Oculina diffusa*, with polyps extended; sea grass and starlet coral in the background, Port Royal, Jamaica.

Plate 2.24. Skeleton of *O. diffusa*, showing widely separated and projecting corallites.

toward the tips of the branches. Even though the corallites are only 2–3 mm across, the septa are distinct since the larger ones protrude above the rims of the corallite walls. The color is yellow brown, but white colonies lacking zooxanthellae may sometimes be found in dark places such as caves. The main habitat of this species, however, is shallow, sheltered sites, 1–10 m deep, where plankton productivity is fairly high and the water sometimes turbid. The species may be common in such places—examples include the mouth of Kingston Harbor in Jamaica, and the north coast of Trinidad.

# ▧ Family Meandrinidae

The family is characterized by skeletal features that are relatively large, solid, and smooth in appearance. The septa in the corallites are clearly visible, even, and widely spaced; however, the family contains a wide diversity of forms among the four species occurring in the Caribbean. Of these, two are meandroid, while the other two have separated corallites; one species is branched, one forms pillars, and the other two form plates and mounds. All four species occur occasionally and sometimes frequently at medium depths (5–20 m) on most Caribbean reefs. Each of the four species is currently placed in a different genus, and all four genera are restricted to the Atlantic. Two genera, *Meandrina* and *Dichocoenia*, are sometimes separated into more than one species on the basis of slight differences in colony and corallite form.

*Dendrogyra cylindrus*, pillar coral (plates 2.25–2.26)

This magnificent coral looks like a castle out of a fairy tale; its tall pillars reach toward the surface from depths of about 5–20 m while fish weave among the turrets. A large colony may be 1.5 m or more in basal diameter and more than 2 m tall. The color of this coral is pale brown, and it appears somewhat fuzzy on account of the tentacles of the polyps, which are extended during the day, obscuring the meandroid morphology. The characteristic colony form of this species makes it impossible to mistake for anything else. An interesting feature is that the upright growth of the pillars can give an indication of past disturbance and subsequent growth rate. If a colony is toppled by waves, but survives, new pillars grow upright from the fallen base like new trunks from a toppled but still rooted tree. If the time of the disturbance is known, the height of the new pillars can be used to estimate growth—usually about 1 cm per year. Pillar coral is not particularly common, but it can be seen regularly on the fore reef of many Caribbean reefs.

Plate 2.25. Pillar coral, *Dendrogyra cylindrus*, off Discovery Bay, Jamaica.

Plate 2.26. Closer view of *D. cylindrus*, showing tentacles extended during the day.

*Meandrina meandrites*, maze coral (plates 2.27–2.28)

The name is no doubt derived from the meandroid arrangement that is similar to, though larger and more obvious than, that of the previous species. The meandering ridges are about 10 mm apart, lined on each side with prominent septa and with rows of mouths in the grooves between the ridges. Maze coral overlaps pillar coral in color, often being pale brown, but the range of browns is greater, both darker and richer shades being found. The structure of the colony is usually a roughly circular, fairly thick plate, often about 20 cm in diameter, firmly attached to the substrate, but mounds and encrusting sheets up to 1 m or more in diameter are sometimes found. Small colonies 5–10 cm in diameter may be flowerlike, the meandroid grooves opening like petals, and such forms have sometimes been named as separate species (e.g., *M. braziliensis*). The polyps of *M. meandrites* are retracted by day, but at night they expand to a forest of thick tentacles entirely obscuring the skeleton beneath (plate 1.10). This species is frequently encountered on reefs, generally below about 5–10 m depth.

*M. meandrites* is of course a type of brain coral but is easily distinguished from other brain corals by its characteristic color and texture. The clean and prominent lines of the thick septa are quite different from those of typical brain corals (e.g., *Diploria* spp.) in which the septa are narrower and more closely packed.

Plate 2.27. Platelike, relatively dark brown colony of maze coral, *Meandrina meandrites*.

Plate 2.28. Pale brown *M. meandrites* forming a mound.

*Dichocoenia stokesi*, elliptical star coral (plates 2.29–2.30)

*D. stokesi* can be found on most coral reefs at depths of 5–20 m, but it is not normally a common coral; however, it appears to tolerate a wide range of habitats. Off Margarita Island in the southern Caribbean, for instance, where, due to upwelling, temperatures are often lower and water more turbid than most corals prefer, it is one of only a handful of common coral species. Colonies grow in the form of regular hemispherical mounds, or less frequently as thick plates, generally very firmly attached and 30 cm or more in diameter. Color is generally a rich brown but may be paler. The large, separate corallites are irregularly rounded, often elliptical, and about 3–5 mm in smallest diameter. The edges of each corallite project slightly above the surrounding skeleton. As with those of other meandrinids, the widely spaced septa are clearly visible when the polyps are retracted; however, polyps are sometimes partially expanded during the day. In many colonies, some of the corallites are elongated to several times their width, in which case there may be more than one mouth per polyp (plate 2.30). Plate-like colonies with smaller rounded corallites have sometimes been separated as a different species: *D. stellaris*.

There are two other corals with which this species could be confused: the faviids *Favia fragum* and *Montastraea cavernosa*, both of which have relatively large projecting corallites; however, in *M. cavernosa* the color of the living coral is much more variable (green, purple, gray, etc.), the tops of the projecting corallites are rounded rather than flat, and when the polyps are contracted the septa cannot be seen clearly. In *F. fragum* the colonies are much smaller and the corallites are usually more closely spaced and do not project as much as those of *D. stokesi*.

Plate 2.29. Elliptical star coral, *Dichocoenia stokesi*, with brown and green algae in the background.

Plate 2.30. Closer view of *D. stokesi,* showing a variety of corallite shapes. The white spots are the tips of tentacles.

*Eusmilia fastigiata,* smooth flower coral (plates 2.31–2.32)

This species was recently transferred from the family Caryophylliidae—indicating the difficulties of assessing higher-order relationships in stony corals. Caryophylliids are mostly solitary cup-corals, and although *Eusmilia* is very clearly colonial, forming branched hemispherical colonies, its form is of many solitary cup-corals, each on the end of a branch. The lower parts of the longer branches are not living, so polyps are separated from their neighbors, although the tentacles of neighboring polyps may touch when extended. The polyps divide as they grow, and polyps in the process of division can usually be seen (plate 2.32). Each polyp is 1–2 cm or more in diameter, and the color ranges from pale brown to green or gray, sometimes with blue or pink tints. Like the other meandrinids, smooth flower coral is occasional to frequent in occurrence rather than common. Most reef habitats 5–20 m in depth and sheltered from wave action can be colonized by this species.

*E. fastigiata* is called smooth flower coral to distinguish it from spiny flower coral (*Mussa angulosa*), but there is no risk of the two being mistaken for each other as the polyps of *M. angulosa*, at 5–8 cm in diameter, are much bigger and fleshier and the septa are spiny.

Plate 2.31. Colony of smooth flower coral, *Eusmilia fastigiata*, surrounded by algae.

Plate 2.32. Close-up of retracted polyps of *E. fastigiata*, showing the prominent septo-costae and recently divided polyps (*top and bottom right*).

## ■ Family Siderastreidae

This family is represented in the Caribbean by one genus, *Siderastrea*. Characteristics include small sunken corallites, 2–4 mm across, shaped like inverted cones containing numerous septa edged with granules and with poorly developed corallite walls. There are two Caribbean species.

*Siderastrea siderea*, massive starlet coral (plates 2.33–2.35)

This is a very common species, ranging from shallow water in lagoons out onto the fore reef and down to 15 or 20 m depth. In shallow water it forms large domed colonies often more than 1 m across, as well as smaller boulders, and encrusting lumps in deeper water. The color is most often a fairly uniform midbrown but ranges from reddish or pinkish brown through to grayish. Because of the depressed corallites, 3–5 mm across, and the even color, the surfaces of colonies look fairly smooth and the septa, although numerous (>40), are not obvious in life. Colonies of this coral are sometimes found apparently growing out of sand, and such colonies may form the basis for development of small coral patches that attract settlement of other species on dead and eroded parts of the original colony.

Plate 2.33. Large colony of massive starlet coral, *Siderastrea siderea*, with yellow sponge, off Tobago, at depth of about 8 m.

Plate 2.34. Closer view of *S. siderea*.

Plate 2.35. Close-up of retracted *S. siderea* polyps.

*Siderastrea radians*, lesser starlet coral (plate 2.36)

This common coral forms smaller colonies than *S. siderea* and it also generally has smaller corallites (2–3 mm across), similarly depressed but with paler margins and darker centers to the polyps, against which the white edges of the primary septa are clearly visible. Septa are numerous but fewer (<40) than in *S. siderea*. These corals are pale brown with dark stars marking the corallite centers. It is a species with extreme tolerance of bright light and warm water and is often the only coral to be found in very shallow lagoons, where it encrusts rock, stones, or sometimes living

Plate 2.36. Small colony of lesser starlet coral, *Sideastrea radians*, encrusting reef rock.

mollusk shells. It may also form small balls of coral, rolling free of any attachment.

One other species—*S. stellata*—has been described and appears similar to *S. radians*. Whether it is a true species or a variety remains to be determined.

## ■ Family Agariciidae

This family is represented in the Caribbean by the whole of the genus *Agaricia* and one species of the genus *Leptoseris*. The family is similar to Siderastreidae in having sunken corallites, but there are relatively fewer septa and these run (as septo-costae) between corallites. The colony form of the Caribbean agariciid species is generally flattened and platelike, the plates being very thin, often less than 5 mm thick, and tapering to sharp edges. The plates are attached at a single point below, or along an edge, or encrusting. The corallites themselves are usually arranged in single rows within grooves. These grooves or valleys run roughly parallel to each other, separated by ridges, and are arranged in roughly concentric circles around the center of growth of the flattened, circular colony, parallel to the growing edge. The often elegant shapes of these corals, and the surface textures produced by the ridge and groove systems, make agariciid colonies particularly beautiful features of the deeper reefs where many of the larger species live.

Members of the genus *Agaricia* are often difficult to distinguish from each other, especially underwater, because several of the species are very similar in overall form and in superficial detail. Corallite size and spacing and the dimensions and regularity of the ridge and groove systems vary within species and may overlap between species; thus the identification of any particular colony may depend on a combination of features. Molecular genetic evidence is helping to establish relationships within the genus (Stake 2007).

*Agaricia agaricites* (syn. *Undaria agaricites*), lettuce coral (plates 2.37–2.40)

This is the commonest species of the genus, and one of the commonest species of corals in the Caribbean. It has very wide habitat tolerance, shallow to deep, turbid to clear, and extends onto shaded cliffs and into crevices. It also shows a wide range of colony form: although normally somewhat flattened, it can form irregularly angular lumps, vertically arranged plates, horizontal plates, or mixtures of these forms. A frequent mixture is an encrusting plate from which arise various upright plates and lumps. Upright plates have corallites on both sides, but horizontal and encrusting plates lack corallites on the underside. Compared with those of other Caribbean agariciids, the central regions of plates can be relatively thick (1–4 cm) and sturdy. Colony size is generally modest, rarely more than 30 cm in diameter, and very small colonies a few centimeters across may be common.

The corallites are small with a deep central pit and weak columella and arranged in grooves, with sharp-edged ridges 2–6 mm apart separating adjacent grooves. The grooves themselves are usually fairly short but can be up to 50 mm long with 4–7 corallites per centimeter. Grooves and ridges are only roughly parallel and concentrically arranged, but short grooves with cross ridges often give a somewhat reticulate pattern. The color is highly variable: shades of brown, gray, and green are common, often with highlights of purple, green, or orange.

Some of the extremes of variability in this species have been elevated to species status by some authorities. One such possible species is *A. humilis*, in which the grooves are shortened to pits, each containing one or a very few corallites (plate 2.39). Different growth forms (upright plates, horizontal plates, shorter or longer grooves) may be found growing next to each other so seem not to represent responses to local environmental conditions. Molecular evidence suggests that most of these different growth forms are varieties of *A. agaricites*, with the possible exception of *A. humilis*, which may deserve species status (Stake 2007).

Plate 2.37. Lettuce coral, *Agaricia agaricites*, with upright plates; gorgonian and star coral in the background.

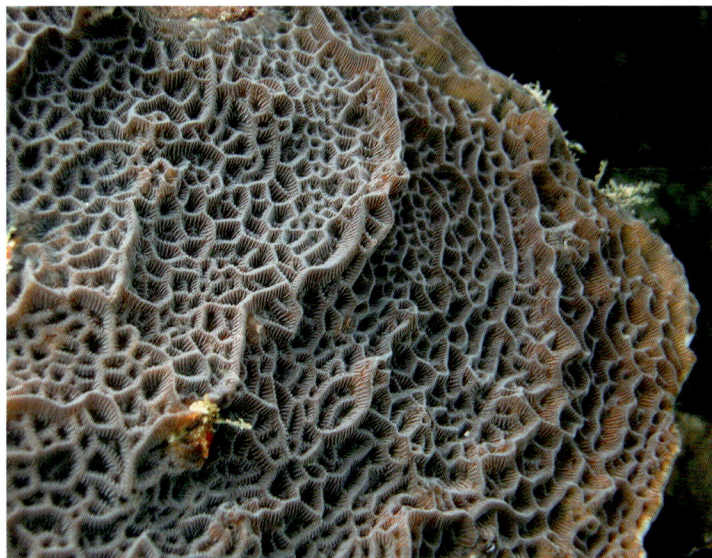

Plate 2.38. Upward-facing plate of *A. agaricites*, showing reticulate pattern of ridges.

Plate 2.39. Encrusting colonies of *A. agaricites*, showing different groove and ridge patterns; the colony on the left with very short grooves may be *A. agaricites humilis*.

Plate 2.40. Skeleton of *A. agaricites*, showing sharp ridges and corallites in deep grooves; septo-costae run between corallites over the crests of ridges.

*Agaricia tenuifolia* (syn. *Undaria tenuifolia)*, thin leaf lettuce coral (plate 2.41)

Thought by some to be a variety of *A. agaricites*, this coral is an extreme exponent of the upright-plate growth form. Unlike those of *A. agaricites*, however, the plates of *A. tenuifolia* are more

Plate 2.41. Thin leaf lettuce coral, *Agaricia tenuifolia*, on the Pedro Bank, south of Jamaica.

numerous, thinner, sometimes leaflike, and easily broken. The upright plates are individually quite small—less than 10 cm across—but may be branched and elongated. They can be flat but more often are curved into vertical folds and corrugations. Rows of corallites are found on both sides of the plates, situated in irregular grooves separated by ridges; the arrangement is similar to that in *A. agaricites*, with both grooves and ridges running roughly parallel to the growing edge of the plate. Color is in shades of brown, tending to yellow and green.

Despite the small size of each plate, whole colonies of many thin upright plates may be very wide, and in some places this coral can form extensive carpets, dominating the reef from shallow water to about 8 m deep. This carpeting occurs in the sheltered waters of the Bocas del Toro in Panama. In more exposed reef-top conditions, however, colonies are smaller and plates shorter and thicker. On the Mesoamerican Barrier Reef this coral has been described as a pioneer species, often settling in large numbers on dead and broken coral rubble after hurricanes; however, abundance appears very dependent on geographic location, *A. tenuifolia* being common in some areas but rare or absent in others.

*Agaricia fragilis*, fragile saucer coral (plate 2.42)

This is an infrequently encountered species with small flat or saucer-shaped colonies up to 15 cm across and less than 5 mm thick. Colonies grow in shaded, sheltered locations over a range of depths from about 3 m downward; possibly their fragility precludes more exposed locations. The species is characterized by having very closely spaced and sometimes low-relief ridges, 2–4 mm apart. The corallites are also very small, with a weak columella, and occur only on the upper side of the colony, 5–8 per centimeter along the grooves. Grooves and ridges are relatively long compared with those of *A. agaricites*, but the live color is similar in shades of brown, gray, and green.

Plate 2.42. Fragile saucer coral, *Agaricia fragilis* (*left*), showing narrow grooves and ridges in contrast to sunray lettuce coral, *Leptoseris cucullata* (*right*), with much larger grooves and ridges.

*Agaricia lamarcki*, whitestar sheet coral (plates 2.43–2.44)

The next three species share features of overall colony morphology, all growing as large thin plates up to 40 cm or more in diameter, with long, relatively regular ridge and valley systems on the upper side only. In all cases the plates may overlap and curve to form broad shingles, scrolls, and bowls. All three species tend to be found most frequently in relatively deep water (>20 m and down to 50 m or more), and all share a range of colors, including browns, grays, and greens.

Plate 2.43. Two colonies of whitestar sheet coral, *Agaricia lamarcki*.

The commonest of the three species, *A. lamarcki*, is also the easiest to recognize since it has relatively large polyps with white mouths, spaced 3–5 per centimeter along the grooves, which show up as white stars at the positions of each corallite center. The ridges are spaced 4–8 mm apart and have smooth rounded crests (unlike *A. agaricites*, which has sharp-edged ridges). The corallites have a well-developed columella, and the septo-costae are often thinner than the spaces between them and may alternate in length, height, and thickness. Like *A. agaricites*, with which it is often found, *A. lamarcki* tolerates a wide range of environmental conditions and although commoner in deeper water, it may sometimes be found as shallow as 5–10 m.

Plate 2.44. Closer view of *A. lamarcki*, showing long, smoothly rounded ridges with white-mouthed polyps in shallow grooves.

Plate 2.45. Large plates of dimpled sheet coral, *Agaricia grahamae*, at a depth of about 30 m, Runaway Bay, Jamaica.

Plate 2.46. Closer view of the long ridges and grooves of *A. grahamae.*

*Agaricia grahamae,* dimpled sheet coral (plates 2.45–2.46)

This coral is another large and beautiful species, generally found in deeper water. Its proportions are similar to those of *A. lamarcki,* but it differs in lacking white polyps and in having sharper ridges, deeper valleys, corallites that appear deeper ("dimpled"), and septo-costae that rarely alternate and are normally thicker than the spaces between them. Some of these characteristics make *A. grahamae* similar to the flat platelike form of *A. agaricites,* but it differs from the latter in its thinner plates and its longer, more regular ridges and grooves.

*Agaricia undata*, scroll coral (plates 2.47–2.48)

The plates of this coral often form large and elegant bowls and open spirals in deeper water. As in the previous two species the plates are thin, and both grooves and ridges are relatively long and parallel to the growing edge; however, the corallites of *A. undata* are very small, 5–8 per centimeter, and near the growing edge they occur tucked in at the base of the steep outer-facing slopes of the rounded ridges, facing out across rather flat valleys (outward-facing corallites at the edges of colonies are found

Plate 2.47. Bowl-shaped colony of scroll coral, *Agaricia undata*.

Plate 2.48. Closer view of edge of an *A. undata* plate, showing small, outward-facing corallites and wide flat grooves.

in other *Agaricia* species, but this condition is most marked in *A. undata*). Toward the centers of large colonies the corallites are placed more centrally in the grooves between rounded ridges about 6–8 mm apart. Colonies are generally pale brown in color. *A. undata* and *A. grahamae* are generally described as uncommon, but this perception may be due partly to the difficulties of distinguishing them from other more familiar species, and to their less easily accessible, deeper habitat.

*Leptoseris cucullata* (syn. *Helioseris cucullata*), sunray lettuce coral (plates 2.49–2.51)

This species, while superficially similar to the other Caribbean agariciids, is different in detail and relatively easy to recognize. It is also quite common from about 5 m downward to 30 m or more and is tolerant of a wide range of environmental conditions. The colonies usually grow as small thin plates up to about 30 cm across, often closely following the underlying substrate and attached only at a single point.

The corallites occur only on the upper surface, are larger than those of other agariciids, and lack a columella (plate 2.51). They are angled strongly outward, facing toward the edge of the colony.

Plate 2.49. Small colony of sunray lettuce coral, *Leptoseris cucullata*.

Plate 2.50. View from the edge of a plate of *L. cucullata*, showing irregular ridges and green outward-facing polyps.

Plate 2.51. Skeleton of *L. cucullata*, showing septo-costae running between corallites over the crests of the ridges.

Large irregular ridges parallel to the colony margin and 4–10 mm apart are formed from rows of these outward-facing corallites, about 3 per centimeter. Live colonies are usually some shade of brown, often a rich rust brown, with prominent pale septo-costae running between corallites over the crests of the ridges. The polyps nestled in the corallites at the base of the ridges often show bright orange, yellow, or green spots and splashes.

## Family Dendrophylliidae

This is a family of mostly azooxanthellate corals, including many cup-corals; however, the family also contains many large colonial zooxanthellate species in the Indo-Pacific. A feature of the family is that the corallites have porous walls. The Caribbean shallow-water species of this family are azooxanthellate and mostly inconspicuous cup-corals, but one is colonial and rather noticeable.

*Tubastraea coccinea*, orange cup-coral (plates 2.52–2.53)

With pink to orange polyps and bright orange tentacles, this species stands out from its surroundings, especially at night when the tentacles are fully extended. It is colonial and forms irregular

Plate 2.52. Colonies of orange cup-coral, *Tubastraea coccinea*, with tentacles partly extended in turbulent water beneath a coral ledge at a depth of 1–2 m.

Plate 2.53. Closer view of retracted polyps of *T. coccinea* on a reef at a depth of about 25 m.

colonies, often of just a few polyps that project from the basal skeletal mass. Like other azooxanthellate corals, *T. coccinea* is found in environments where light is reduced, but both deep and shallow sites with very different degrees of water movement can be colonized. Elongated corallites, 2 or more centimeters long and on narrow bases, tend to be found in deeper sheltered sites, whereas shorter corallites (<1 cm long) arising from encrusting bases can occur under very shallow reef ledges exposed to strong wave-surge (compare plates 2.52 and 2.53). Corallites vary in diameter from a few millimeters up to about 1 cm.

The species is unusual in that it is an immigrant to the Caribbean from the Indo-Pacific, where it is widespread, probably arriving in the 1930s or 1940s. It is now found throughout the Caribbean and can be fairly common but local in distribution.

## ■ Family Caryophylliidae

This is another family of mainly azooxanthellate cup-corals, non reef builders that live in a wide range of environments, including the deep sea and temperate seas. Molecular genetic studies suggest that the family is highly polyphyletic and in need of revision (Kitahara et al. 2010). Several species currently included in the Caryophylliidae live in shallow tropical seas, where they generally occur in cryptic locations such as beneath boulders, in crevices, and in caves. There may be five or more shallow-water species of such cup-corals in the Caribbean, but only one of these is common or noticeable enough to warrant description here.

Plate 2.54. Partly expanded polyps of hidden cup-coral, *Phyllangia americana*, beneath an overhang in Trinidad at a depth of about 5 m.

Plate 2.55. Fully expanded polyps of *P. americana*, growing in the dark, at a depth of about 15 m, on the walls of a blue hole off Andros Island, Bahamas.

*Phyllangia americana*, hidden cup-coral (plates 2.54–2.55)

The corallites of this species rise individually from the substrate; they are up to about 1 cm in diameter and are short and fat when compared with some other cup-corals. When the polyps are contracted, the septa project above the mouth and show up as short white lines radially arranged around the darker center. The 6 primary septa are especially evident. When the polyps are extended, the long tentacles and swollen mouth region obscure the septa. The color of the polyps is rather variable, but usually includes shades of brown. Polyps are generally unconnected to their neighbors so are said to be solitary, although where one is found, others are almost always close by. In favorable environments, however, groups of polyps may form small irregular colonies connected by skeleton and tissue.

*P. americana* is found in places where light is dim, such as beneath overhangs and in caves. It can be extremely common, studding the dark walls of some Bahamian blue holes where currents stream past.

## ▨ Family Mussidae

Members of this family usually have large fleshy polyps, and often the projecting parts of the septa and corallite margins bear prominent teeth or spines. These features make the corals feel spiny to the touch and explain the common name of cactus coral for several species. Most members of the family are colonial, but there are a few solitary species. Colonies may be massive, encrusting, platelike, or branched, and the polyps may be distinct or joined in meandroid patterns. Molecular genetic studies suggest that the Atlantic mussids are more closely related to each other than to the Indo-Pacific mussids, and the family as a whole is probably polyphyletic (Kitahara et al. 2010). There are eight Caribbean species currently referred to four genera. These are presented below in order of increasing departure from the solitary polyp form.

*Scolymia cubensis,* artichoke coral (plates 2.56–2.58)

This species belongs to a genus requiring reorganization since the other two described species of *Scolymia* live in the Indo-Pacific and, according to molecular evidence, are not closely related to *S. cubensis.* Artichoke corals are solitary corals with large polyps up to 6 cm or more across. Their appearance in life is as large, rough, and often colorful buttons with a fleshy rim. Each is shaped like a shallow saucer with a short attachment stalk below, vaguely reminiscent of the heart of a globe artichoke. The color ranges from yellow green to red, often with spots or patterns of different colors; radial and concentric patterns are common. Close inspection of contracted polyps reveals the radiating pattern of the skeleton since the tips of spines arising from the numerous septa are represented by radiating rows of bumps on the surface of the tissue. The septa are very numerous: 20 or more larger septa have 3 or more smaller septa arising between them, giving perhaps 80–100 septa in total. The spines on the septa are thin and delicate, especially on the smaller septa, where they may form an irregular lattice rather than a series of points (plate 2.58).

Plate 2.56. Group of bright green artichoke corals, *Scolymia cubensis*.

Plate 2.57. Closer view of a red *S. cubensis* polyp.

Plate 2.58. Skeleton of *S. cubensis*, showing very numerous spiny septa.

*S. cubensis* corals can be quite common, especially on reefs subject to some turbidity. Being individually small, however, they are often overlooked. They generally occur below 10 m depth and are not usually found in brightly lit water, but where there is shelter and shade they may occur in 3–5 m depth.

*Mussa angulosa*, spiny flower coral (plates 2.59–2.61)

This species resembles several artichoke corals joined together into a branched colony. In life, the colony looks like a fleshy lump, with each large polyp occupying the end of a single branch and being shaped to fit with its neighbors. Polyps of *M. angulosa* are generally larger (5–8 cm in diameter) and fleshier than those of *S. cubensis* and have a more robust skeleton that extends to the septa, where the spines are better described as teeth, flattened and triangular in outline when viewed from the side. The colors of living polyps are generally duller than those of *S. cubensis*, ranging from green to purple and including blues and grays, often with blotches of lighter colors; however, livid red (presumably fluorescent) is also sometimes seen. Although the species is normally colonial, a solitary form of *M. angulosa* exists and was previously placed in the genus *Scolymia* as *S. lacera*.

Plate 2.59. Gray-blue colony of spiny flower coral, *Mussa angulosa*, with pink highlights.

Plate 2.60. Brown-pink colony of *M. angulosa*, surrounded by green algae.

Plate 2.61. Solitary polyp of *M. angulosa* (previously called *Scolymia lacera*) in a silty environment at a depth of about 20 m.

*M. angulosa* is an occasional or rare coral; like *S. cubensis*, it is generally found in water deeper than about 10 m in shaded and sheltered environments and in places subject to some turbidity. It is the sole member of its genus and found only in the Caribbean.

### *Isophyllia rigida* (syn. *Isophyllastrea rigida*), rough star coral (plate 2.62)

This coral forms small mounds or lumps, up to 20 cm across, studded with rather large, mainly separate polyps, retracted by day into corallites 10–15 mm across. The corallites are irregular and have spiny rims formed from corallite walls with projecting spiny

Plate 2.62. Colony of rough star coral, *Isophyllia rigida*, surrounded by a variety of algae.

septa. In life, the spines are concealed beneath the fleshy polyp tissues. The corallite walls form ridges with a groove along the top created by gaps between the outer ends of the septa of adjacent polyps. The color of colonies ranges through shades of green, from yellowish through grayish; in many cases the centers of polyps are a different color from that of the surrounding ridges.

The species is found at medium depths of 5–15 m but is generally rather rare. It was previously classified under a separate genus—*Isophyllastrea*—but there appears to be insufficient evidence to justify this classification (Fenner 1993). As it stands, the genus *Isophyllia* contains only two species, this one and the next, both occurring only in the Caribbean region.

Plate 2.63. Colony of sinuous cactus coral, *Isophyllia sinuosa*, with a green sea urchin, *Echinometra viridis*.

Plate 2.64. Closer view of *I. sinuosa*, showing the shortened meandroid arrangement of polyps.

*Isophyllia sinuosa*, sinuous cactus coral (plates 2.63–2.64)

This species is very similar to *I. rigida* and differs mainly by the meandroid form of its corallites. The polyps are situated in rows within short meandering valleys between ridges formed, as in *I. rigida*, from the corallite walls and spiny septa. The number of polyps per valley ranges from 1 to about 5, with valley width being 15–20 mm. The overall colony form is of a small (10–20 cm diameter) low mound. Color is generally in shades of green.

Like the previous species, this coral is found at moderate depths on the reef and is generally rare. It may sometimes grow less as a mound and more as an encrusting sheet, in which form it may be confused with species of *Mycetophyllia* (described below); however, the valleys of *I. sinuosa* are normally shorter and narrower than those of *Mycetophyllia*.

### Genus *Mycetophyllia*

The genus *Mycetophyllia* is entirely Caribbean, with four (possibly five) presently recognized species. In all cases the colonies grow as flattish plates. Young colonies are initially attached by just a central short stalk, but as they grow they may form additional attachments. The polyps are borne only on the upper surface, and in all but one species they are situated in flat-bottomed valleys between spiny ridges that penetrate in from the colony edge toward the center. This initially radial arrangement of the ridges becomes increasingly irregular and sinuous in larger colonies. As in *Isophyllia*, the ridges consist of adjacent upturned corallite walls with their projecting spiny septa. In life the ridges are rounded in cross-section and covered by fleshy tissue. The tentacles are located on the ridges and in their retracted state during the day contribute to the fleshy appearance of the ridges.

*Mycetophyllia lamarckiana* (syn. *M. danaana*), ridged cactus coral (plates 2.65–2.68)

This is the commonest species of the genus, occurring at moderate depths (10–20 m) or shallower in shaded locations. Colonies tend to be shaped like plates or domes, usually 10–20 cm in diameter. The ridges vary in number and in how far they penetrate toward the center of a colony. When numerous, they are 1–2 cm apart and often go all the way into and across the centers of plates, the valleys between appearing relatively deep; when the ridges are sparse, however, the valleys appear wide and flat. Ridges may also occur as short sections or knobs. The color is variable: chocolate brown, bright green, speckled gray and white, yellowish, and purple can all be found, often with ridges and valleys of different shades.

Although this is the commonest species of *Mycetophyllia*, it is at best occasional to frequent on Caribbean reefs, and not always present. It appears to tolerate increased turbidity and is often more frequent, and shallower, on slightly turbid reefs.

Plate 2.65. Colony of ridged cactus coral, *Mycetophyllia lamarckiana*, with numerous ridges and narrow valleys.

The species is often split into two: *M. lamarckiana* with sparse ridges and *M. danaana* with numerous ridges; however, these differences seem to be only ones of degree, and many colonies can be found that appear intermediate. In addition, there are other features in which they closely resemble each other. For instance, in both forms the septa are thinner than the gaps between them and bear delicate spines; the abundance and habitat are also similar. My conclusion, therefore, supported by communication with other scientists, is that the two represent a single variable species.

Plate 2.66. Colony of *M. lamarckiana* with few ridges and wide valleys.

Very young colonies can be confused with *Scolymia cubensis*, since at just a few centimeters across, colonies of *M. lamarckiana* may have only one fully developed polyp and have not yet developed ridges; however, the edges of such small colonies are often scalloped, indicating where ridges will form, and new polyp development can sometimes be seen as irregularities in the septa. In addition, the septa are less numerous, less closely packed, and less elaborately spiny than those of *S. cubensis*.

Plate 2.67. Closer view of broken ridges and narrow valleys in the center of a large colony of *M. lamarckiana*.

Plate 2.68. Skeleton of *M. lamarckiana*, showing the septa, underlying structure of ridges, and polyp centers in the valleys.

*Mycetophyllia ferox*, rough cactus coral (plates 2.69–2.70)

This species and the next form large, relatively thin plates up to 50 cm in diameter that tend to encrust or mantle the underlying surface. *M. ferox* has numerous ridges running inward from the edges of the colony, often joining or breaking and forming narrow valleys that may close at one or both ends. The ridges themselves are low, roughly square in cross-section, and septa are thinner than the spaces between them. Polyp centers tend to be slightly

Plate 2.69. Green colony of rough cactus coral, *Mycetophyllia ferox*, encrusting the underlying rock.

Plate 2.70. Brown colony of *M. ferox*, showing slightly raised orange polyp centers in the valleys.

raised and may be a different color than the surrounding tissue. The general color of live colonies is variable but often dull: greenish to gray brown, with somewhat contrasting ridges.

The species is occasional or rare and tends to be found in sheltered and low-light environments, either in slightly turbid water or at deeper (15–30 m) and relatively steeply sloping sites.

*Mycetophyllia aliciae*, knobby cactus coral (plates 2.71–2.73)

*M. aliciae* is a rather characteristic species that forms large colonies up to 50 cm in diameter with few and often broken ridges; the septa on the ridges are relatively thick with coarse spines. Polyps are raised and pale in color, contrasting with the darker surrounding tissue. Colonies may take the shape of the underlying substrate or may grow as a rough disk, flat or slightly domed. Its shallowest occurrence tends to be deeper than the preceding species and it extends down further, to more than 50 m. It may be frequent or common in deeper or slightly turbid waters. The color can be quite variable but is usually fairly dark: rich browns and greens are common, with the paler ridges and polyps being clearly distinct.

Plate 2.71. Green colony of knobby cactus coral, *Mycetophyllia aliciae*, showing few ridges and pale, slightly raised polyp centers.

Plate 2.72. Closer view of *M. aliciae*, showing details of the ridges and polyp centers.

Plate 2.73. Young colony of *M. aliciae*, about 5 cm in diameter, showing early stages of ridge formation.

Plate 2.74. Colony of ridgeless cactus coral, *Mycetophyllia reesi*, on an underwater cliff at a depth of about 30 m, Runaway Bay, Jamaica.

*Mycetophyllia reesi*, ridgeless cactus coral (plates 2.74–2.75)

This is the most easily recognized and also the least often encountered species of *Mycetophyllia*. It occurs sparsely in deep water, on steep shaded slopes at 30 m to below 50 m. As its name suggests, it lacks ridges and appears as a smooth, slimy, somewhat undulating plate generally less than 30 cm across and attached in the center or near an edge. Closer inspection reveals that the polyps are arranged in roughly concentric lines, each being raised as a small bump about 5 mm across. These polyps are unusual in that they lack tentacles and feed using a mucus net (Goldberg 2002). The color is generally dark gray with green, blue, or brown tints, sometimes with lighter streaks and blotches.

Occasional specimens of *M. aliciae* may lack ridges, but the surface always has a rough appearance that contrasts with the smoother slimy look of *M. reesi*.

Plate 2.75. Closer view of *M. reesi*, showing rows of polyps.

# Family Faviidae

According to molecular genetic evidence, this large family is highly polyphyletic (Huang et al. 2011). As in the Mussidae, the Atlantic faviids are generally more similar to each other than to the Indo-Pacific faviids. Seven genera currently assigned to this family occur in the Caribbean, of which two (*Montastraea* and *Favia*) are probably themselves polyphyletic since the Atlantic and Pacific representatives are not sufficiently closely related to be grouped in the same genus. The remaining five Caribbean genera are restricted to the Atlantic (*Cladocora, Solenastrea, Diploria, Colpophyllia,* and *Manicina*). One of the latter, *Cladocora*, forms small and inconspicuous branched colonies and is not covered here. All species of the other Caribbean genera form colonies shaped like lumps, mounds, or heads, sometimes of very large size. Some have separate corallites (*Favia, Montastraea,* and *Solenastrea*), whereas in others the corallite pattern is meandroid (the brain corals: *Diploria, Colpophyllia,* and *Manicina*). In molecular genetic analyses, these three brain coral genera and the Atlantic species of *Favia* appear closer to the Atlantic Mussidae than to other faviids, whereas *Solenastrea* appears closer to the Oculinidae (Kitahara et al. 2010).

*Favia fragum*, golfball coral (plate 2.76)

As well as occurring in the Caribbean, *F. fragum* also occurs off the coasts of Brazil and West Africa. It is a small species, growing as a lump or encrustation usually around 5 cm in diameter. It may sometimes entirely encrust small loose objects to form spherical, free-living colonies—hence the common name. The corallites are about 5 mm in diameter and irregularly rounded and separate from each other. New corallites bud from within existing ones (intratentacular budding), and the corallite shapes are thus often distorted from circular to elongated or polygonal. Separation between adjacent corallites is rather variable, from narrow (consisting simply of adjacent corallite walls and septa) to wider, in which a shallow groove is present between the walls, making each corallite appear to project slightly. Septa are numerous (>20) and covered with tiny spines. Both primary (complete) and secondary (incomplete) septa are present.

*F. fragum* is fairly common in shallow water (intertidal to 5 m), but being small, it is easily overlooked. The color of living colonies is generally some shade of brown, often yellowish or greenish brown and frequently with pale patches.

Plate 2.76. Colony of golfball coral, *Favia fragum*, surrounded by algae in a reef-flat environment at a depth of about 1 m.

*Montastraea annularis,* lobed star coral (plates 2.77–2.78)

This is one of the commonest coral species at 5–20 m depth in the Caribbean and is often the dominant coral. It grows as a mass of closely adjacent lobed heads arising as thick branches from a single base. A typical colony might be 1–2 m in diameter and 75 cm high, made up of 10–20 lobes, each 10–30 cm in diameter at the outer end. These outer ends bear the living polyps while the largely dead basal parts of the colony, consisting of clustered columns branching at intervals, provide attachment surfaces and hiding places for numerous small animal and plant species. This coral could therefore be described as forming branched colonies, but the branches are so thick and close together that it tends to be classed as a massive, mound-forming species. Large colonies are often broken by storms such that lobes at the edges fall sideways, but living tissue on the outer ends may remain alive and continue growing. By this means, the same genetic identity may split and grow to cover many square meters of reef space. Because of its large size and high abundance, this species is a major framework builder of Caribbean reefs.

The corallites are fairly small (2–4 mm in diameter; see plate 1.6), discrete and circular; new corallites arise between existing ones (extratentacular budding). There are normally 12 regularly arranged primary septa arching up from the corallite edges so that each corallite protrudes somewhat from the colony surface; small secondary septa occur between the primaries. Color in life is yellow green, green, brown, or occasionally gray.

There has been controversy over many years about variability in this species and whether certain varieties should be given species status. Those advocating several species separated the "*Montastraea annularis* complex" into three: *M. annularis, M. faveolata,* and *M. franksi.* Molecular evidence suggests that *M. franksi* (a form in which the surface is covered irregularly with small lumps and patches of pale tissue) should be retained within *M. annularis;* however, *M. faveolata,* described next, appears to be a distinct

species. Interestingly there seems to be some geographical variation in the genetic relatedness of the three forms, suggesting that
the degree of reproductive isolation varies from region to region
and that in some places hybridization may occur (Fukami et al.
2004a).

Plate 2.77. Large colony of lobed star coral, *Montastraea annularis*, with both large and small lobes.

Plate 2.78. Closer view of *M. annularis*, with small lobes.

*Montastraea faveolata*, mountainous star coral (plates 2.79–2.81)

The corallites of this species are indistinguishable in size and shape from those of *M. annularis*. The colony form, however, is as a continuous mound or sheet, not branching to form lobes. The mounds may be very large—several meters high and across—and the surface is usually textured with more or less regular ridges, or rows of lumps, running radially from the colony center. Plate-like growths angled downward occur around the lower edges of colonies and on slopes in deeper water and are often similarly ornamented with rows of lumps.

Plate 2.79. Small (75 cm across) colony of mountainous star coral, *Montastraea faveolata*, showing descending rows of bumps.

Plate 2.80. Closer view of rows of bumps on a colony of *M. faveolata*.

Plate 2.81. Great star coral, *Montastraea cavernosa* (*left*), contrasted with *M. faveolata* (*right*), showing the difference in retracted polyp size.

This species may be as abundant as *M. annularis*, but it appears to have a deeper preferred depth range, continuing below 20 m down to 40 m or more. In life, *M. faveolata* is usually gray, brown, or green but is rarely yellowish.

*Montastraea cavernosa*, great star coral (plates 2.81–2.84)

This is another large common coral mainly in medium to deep reef environments (15–40 m), but small colonies may also occur sporadically in shallow water just a few meters deep. Large colonies grow as mounds, pinnacles, and sheets up to a meter or more high or across, and colonies may also be thickly encrusting. It is easily distinguished by day by its large, separate polyps 6–12 mm

Plate 2.82. Pinnacle-shaped colony of *M. cavernosa*.

Plate 2.83. Contrasting colors of two colonies of *M. cavernosa*.

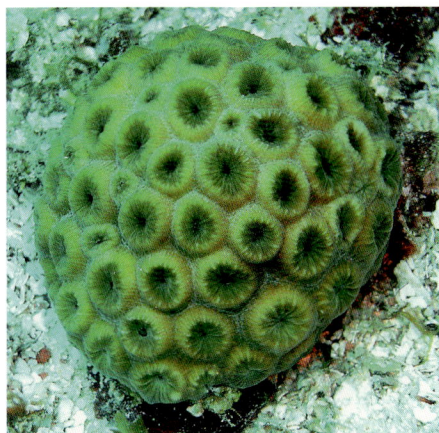

Plate 2.84. Young colony of *M. cavernosa*.

in diameter, which protrude as numerous regular domes with a mouth in a depression at the summit of each. The domed shape of the corallites distinguishes *M. cavernosa* from other Caribbean corals with large corallites, such as *Dichocoenia stokesi*, in which the corallites have flattish tops; by night, however, when the polyps extend their tentacles these two species can appear similar. The color of *M. cavernosa* is often rather dark: browns and greens are common, sometimes flecked with white; occasionally purple or red colonies may be found.

Molecular genetic evidence suggests that this coral is not closely related either to other *Montastraea* species or to other faviids (Kitahara et al. 2010).

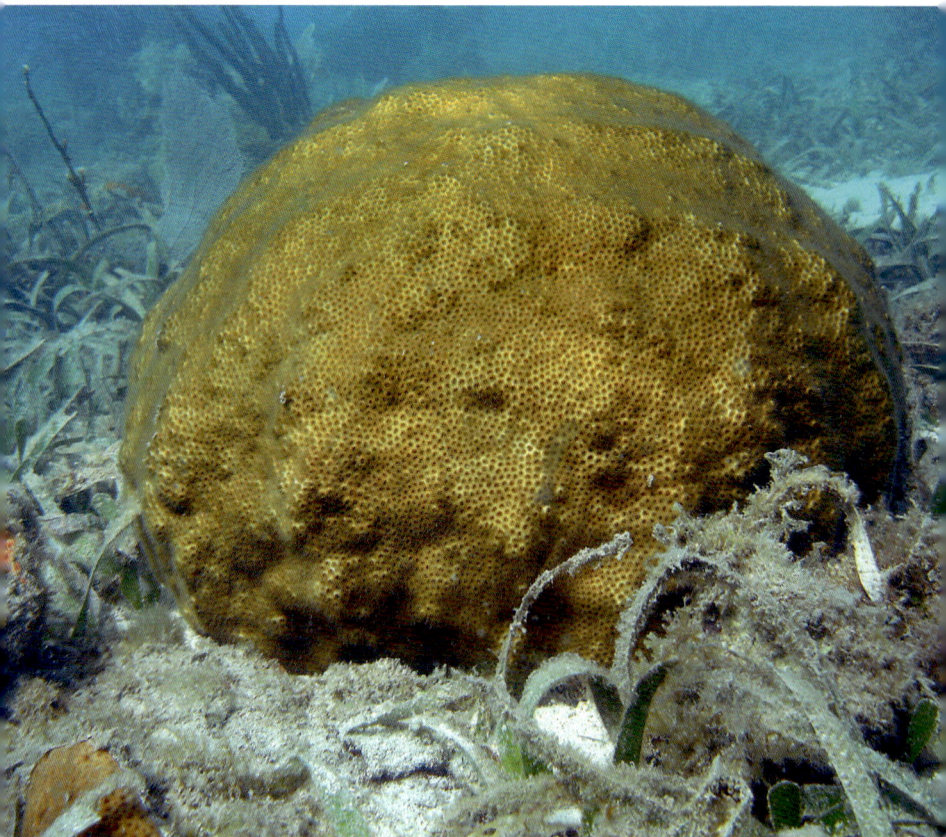

Plate 2.85. Colony of smooth star coral, *Solenastrea bournoni*, among sea grass in the Port Royal Cays, Jamaica.

*Solenastrea bournoni*, smooth star coral (plates 2.85–2.87)

*S. bournoni* can grow into quite large dome-shaped colonies 30–50 cm high and across, and sometimes larger. The outline is generally smooth with minor irregularities. The separate corallites are superficially like those of *Montastraea annularis*, being similar in size (diameter 2–3 mm) and circular in shape; however, the septa protrude less, so the corallites appear lower and flat topped, and have paliform lobes (plate 2.87). In life, the color is yellowish green or yellowish brown and the polyps are often expanded by day, giving the colony a slightly fuzzy look. When contracted, the polyps concentrate their color into the corallites, which then appear darker brown against the lighter surrounding tissues.

*Solenastrea bournoni* is generally uncommon but may be frequent in its preferred environment: shallow (2–5 m), often turbid sites, with sandy or muddy substrates and sea grass.

Plate 2.86. Closer view of *S. bournoni*, showing polyps extended during the day.

Plate 2.87. Close-up of skeletal features of *S. bournoni*, showing slightly projecting corallites with paliform lobes on the septa.

*Solenastrea hyades*, knobby star coral

*Solenastrea hyades* differs mainly from *S. bournoni* by its growth form. Colonies grow as irregular lumps and columns, tending to form lobes or short branches toward the top. The color of the polyps is dark brown so that in life the entire colony appears to be spotted.

This is a rare coral in the Caribbean but becomes commoner further north, where it tolerates the lower temperatures found off the southeast coast of the United States, as far north as South Carolina. It occurs in a broader range of environments than *S. bournoni*, being found from shallow turbid sites down to deeper (20 m) reef sites.

## Genus *Diploria*

This genus contains most of the Caribbean brain corals. The whole surface of brain corals is textured by meandering ridges and valleys reminiscent of the outside of a human brain. The valleys contain a series of mouths, with the tentacles extending from the valley walls, where they may sometimes be detected half-extended by day. The ridges are formed from the rows of septa radiating from adjacent valleys. This meandroid morphology is generally very obvious, and in two of the three species of *Diploria* the commonest overall form is rounded, resembling a huge brain in both shape and surface appearance.

*Diploria strigosa*, symmetrical brain coral (plates 2.88–2.89)

This is the commonest brain coral in the Caribbean, occurring mostly in shallow water (1–10 m), where it may be the dominant coral. It also extends down to the deeper reefs (20–30 m) but is much less common there. It forms large (1 m or larger) rounded colonies in shallow water, and low mounds or encrustations in deeper or shaded locations. The color of living colonies is generally some shade of brown, tending to yellowish, greenish, or gray.

The characteristics by which *D. strigosa* is distinguished from other species of brain coral include the smoothly rounded overall shape, the narrow valleys and ridges—generally about 6–9 mm from ridge crest to ridge crest—and the absence of a groove along the crest of the ridges (although a fine line may be present).

Plate 2.88. Large colonies of symmetrical brain coral, *Diploria strigosa*, at a depth of 1–2 m.

Plate 2.89. Close view of the meandroid corallite pattern of *D. strigosa*.

*Diploria clivosa*, knobby brain coral (plates 2.90–2.91)

As its common name suggests, this brain coral does not show smooth, rounded curves but develops irregular knobs and lumps on its surface. Although sometimes forming large irregular heads, it more often forms low mounds or encrustations. Valleys and ridges tend to be slightly narrower (4–6 mm between adjacent ridges) than in *D. strigosa*, and the crests of the ridges sharper. The color in life is gray brown or gray green.

*D. clivosa* is not as common as *D. strigosa* but can often be found in shallow water among colonies of the latter; it also extends into very shallow wave-washed environments.

Plate 2.90. Colony of knobby brain coral, *Diploria clivosa*, at depth of 1 m.

Plate 2.91. Closer view of *D. clivosa*, showing surface features; tips of tentacles can be seen on the sides of the ridges.

*Diploria labyrinthiformis*, grooved brain coral (plates 2.92–2.93)

This species is not generally found in very shallow water, occurring mostly in the depth band from 5 to 20 m. While not abundant, it can be found frequently on most Caribbean reefs. Like *D. strigosa*, it grows into smoothly rounded domes, sometimes into nearly perfect three-quarter globes. Its distinguishing characteristic lies in the shape of its ridges, which would be square in cross-section were it not for a conspicuous groove running along the top of each. The cross-section of a ridge is therefore M-shaped; however, these grooves in the ridges are not as deep as the valleys between the ridges where the polyps are. The ridge and valley system is similar in overall size to that of *D. strigosa*, but in *D. labyrinthiformis* the ridges are wider and the valleys narrower. The overall color is pale brown, often with a darker or lighter color in the valleys.

Plate 2.92. Spherical colony of grooved brain coral, *Diploria labyrinthiformis*.

Plate 2.93. Close view of contrasting colors of ridges and valleys in *D. labyrinthiformis*, showing the groove in the middle of the pink ridges and tentacle tips along the sides of the green valleys.

*Colpophyllia natans*, boulder brain coral (plates 2.94–2.96)

*Colpophyllia* is an entirely Caribbean genus with probably only one species. It often grows larger than other brain corals but is otherwise similar in morphology, forming domed colonies or thick crusts with a smooth rounded outline; however, its ridges and valleys are at least double the size of those of *Diploria*. This is not always obvious in isolation but becomes so when a colony of *Diploria* is nearby or adjacent (see plate 1.16). In addition, the ridges have a fine groove running along the top. *C. natans* can be common on reefs at depths of 5–40 m. It shows a wide range of

Plate 2.94. Large colony of boulder brain coral, *Colpophyllia natans*, in contrasting shades of green.

Plate 2.95. Closer view of a brown colony of *C. natans* with short valleys.

Plate 2.96. Close view of *C. natans* with wide white valleys and green ridges.

colors, including browns, greens, blues, and yellows, and often the ridges and valleys are of contrasting hues or colors.

A variety of *C. natans* with short valleys containing one or just a few polyps is sometimes found and has been considered by some to be an additional species—*C. breviserialis*.

### *Manicina areolata*, rose coral (plates 2.97–2.98)

Unlike most other Caribbean brain corals, the rose coral is a small species, seldom reaching more than 10 cm in diameter. Small colonies tend to be oval with a single sinuous valley about 20 mm from ridge to ridge, but when fully grown become circular and domed with a more clearly meandroid valley system. The ridges and valleys are only slightly smaller than those of *Colpophyllia natans*,

Plate 2.97. Brown colony of rose coral, *Manicina areolata*, with pale ridges, in a silty environment.

Plate 2.98. Colony of *M. areolata* with brown and green ridges and white valleys.

and as with that species, there is a fine groove along the tops of the ridges, but since the colonies of *M. areolata* are much smaller, there is usually little difficulty in distinguishing the two species. Further differences are found in the details of the skeleton, with the septa of *M. areolata* being more crowded while the columella has a spongy appearance.

Rose coral colonies are either free-living or only lightly attached by a short stalk. They may be found quite commonly on rubble or in sea grass beds from a few meters deep to about 15 m. They are most common in areas with some turbidity. Color is very variable, from browns and greens to purple, blue, yellow, and white, usually with contrasting ridges and valleys and often with regularly arranged lines and streaks of different colors.

## ■ Family Poritidae

This is a large and widespread family comprising five genera, one of which—*Porites*—contains species that occur in the Caribbean. An important family characteristic is the very porous nature of the corallite walls and septa.

### Genus *Porites*

About 50 species are included in this genus, but only four (or six according to some authorities) are found in the Caribbean. These all have closely spaced, small, porous corallites (1–2 mm in diameter; plate 2.99) with a rather conservative pattern of 12 septa, some of which are fused. Details of these corallites, including fusion pattern, denticles, and pali, are sometimes used to distinguish species.

Plate 2.99. Close-up of characteristic skeletal features of *Porites* corals (here *P. porites*). Closely packed corallites are formed from the porous skeleton; corallites have 12 septa, often fused in pairs and bearing pali.

*Porites astreoides*, mustard hill coral (plates 2.100–2.101)

Colonies form smooth or lumpy heads in shallow water (1–10 m) and smooth or lumpy plates or encrustations in deeper reef locations (10–30 m). The common name best applies to the lumpy heads in shallow water, which are mostly about 10–20 cm in diameter and often of a mustard yellow color. In deeper water, plates and encrustations tend to be greenish, brown, or gray and are frequently more than 30 cm in diameter. Plates grow from an attachment at either a central or an edge position, and corallites only occur on the upper side. Polyps are often expanded by day, giving a fuzzy outline to the corals. This is an abundant and widely distributed species, also occurring off the coasts of both West Africa and Brazil.

Plate 2.100. Lumpy massive colony of mustard hill coral, *Porites astreoides*, polyps extended, surrounded by massive starlet coral, *Siderastrea siderea*, at depth of 1–2 m.

Plate 2.101. Large greenish colony of *P. astreoides*, consisting of flattened and overlapping plates, at a depth of about 10 m.

*Porites colonensis*, honeycomb plate coral

This species resembles in some respects the platelike form of *Porites astreoides*. It was described in 1990 from the Caribbean coast of Panama, where it was said to be common, but it is unclear whether it has a wider distribution. It differs from plates of *P. astreoides* in being foliaceous and darker colored (dark brown or red) with contrasting white or green polyp centers. In addition, the pali on the septa are better developed. Molecular studies indicate distinct differences between *P. colonensis* and other *Porites* species, including *P. astreoides* (Stake 2007).

*Porites branneri,* blue crust coral

It is unusual for corals to be blue in color, but this small species is usually blue to purple. It inhabits shallow back-reef environments, where it forms small lumps or knobby encrustations. The lumps or knobs are just a few cm across with the basal encrustations being larger. This species is said to be uncommon, but it may simply be overlooked on account of its small size and its habitat among shallow and overgrown reef boulders and rubble. In this habitat it overlaps small lumpy colonies of *Porites porites,* but this species (described below) is normally brownish or gray in color and has better-developed septa and pali. *P. branneri* is found in the Caribbean and also off Brazil.


*Porites porites,* finger coral (plates 2.102–2.105)

Finger coral occurs in at least three forms. All grow as colonies of outwardly directed branches. Single colonies or groups of colonies can sometimes grow over large areas 2 m² or more in extent. This may occur through breakage of outer branches followed by growth of the fragments and extension sideways to form a thicket. Provided the reef area is not exposed to strong wave action, finger coral is usually abundant and sometimes dominant in shallow water less than 1 m deep to 20 m or more.

The three forms differ mainly in the dimensions of their branches, and some regard them as separate species: *P. porites, P. furcata,* and *P. divaricata;* however, there are many intermediates. Of the three forms, *P. porites* (*porites*) has the thickest branches (2–4 cm thick), and these are often expanded at their tips so they fit against one another at the surface of a colony with little space between (plate 2.102). The branches of *P. porites* (*furcata*) are 1–2 cm thick and often ramifying, leaving spaces between adjacent branch tips. *P. porites* (*divaricata*) has narrow branches less than 1 cm in thickness; it may form dense thickets and appears to favor shallow sites with some turbidity. All three forms share a range of colors, including brown, gray, and green, and all three usually

Plate 2.102. Colony of *Porites porites* (*porites*) with thick branches and polyps extended.

extend polyps by day. *P. porites* (*porites*) is found off West Africa and Bermuda while the two other forms are restricted to the Caribbean. Despite the sometimes different morphologies, molecular studies have so far not indicated sufficient differences to justify dividing this variable group of Caribbean finger corals into more than one species (Stake 2007).

Plate 2.103. Close-up of extended polyps of *P. porites*.

Plate 2.104. Section of a large colony of *P. porites* (*furcata*) with medium branches and polyps extended.

Plate 2.105. Gray-lilac, thin-branched *P. porites* (*divaricata*) among thicker-branched green *P. porites* (*furcata*), polyps mostly retracted.

# 3 | The Ecology of Caribbean Coral Reefs

## Conditions Necessary for the Growth of Reef-Building Corals

Corals require a hard surface (rock or dead coral) to attach to, and they grow best in warm, clear, clean, well-lit, and well-oxygenated seawater. The preferred temperature range in the Caribbean region is fairly narrow: between 20 and 30 °C, with the cooler temperatures found off Bermuda, Florida, and the Bahamas in winter while the warmer temperatures occur further south in late summer. Outside this temperature range, Caribbean corals become stressed and may die; warmer temperatures in particular may lead to coral bleaching (discussed in chapter 4).

Reef-building corals can be found living from the surface of the sea down to about 60 m deep in clear water; however, sufficient coral growth to build reefs is normally found only in the top 20 m or so, where there is enough light to make full use of the photosynthesis of the zooxanthellae. Clear water is important because turbidity reduces the penetration of light, so the depth at which corals can survive in turbid water is reduced. Further, turbidity is usually caused by suspended sediment that can settle on the corals and bury them, so although many corals can grow in slightly or moderately turbid water at shallow depths, coral species richness is severely reduced in very turbid water because only those species that possess efficient sediment-removal systems can survive. As well as often being intolerant of sediment, corals also have little ability to tolerate reduced salinity. High turbidity caused by suspended sediment combined with reduced salinity at the surface is found near the mouths of rivers, especially in the rainy season when floods can turn the sea brown with muddy river water for miles around. For these reasons, coral reefs do not generally develop in estuaries or near the mouths of large rivers.

As well as bringing sediments, rivers also bring various plant nutrients (fertilizers) to coastal waters such as nitrates and phosphates, and organic matter (which releases nutrients during decomposition). Tropical oceanic waters are naturally very low in nutrients, and corals appear to do best in low-nutrient conditions. Raised nutrient levels stimulate the growth of algae, both in the water and on the reef. Extra planktonic algae increase turbidity—reducing light penetration—and provide extra food for coral competitors such as sponges and for organisms that bore into coral skeletons such as certain bivalve mollusks and worms, weakening coral skeletons and causing bioerosion. Extra algae growing on the reef can smother corals and prevent coral larvae from settling. So, for several reasons, raised nutrient levels reduce both the abundance and the species richness of corals.

The requirement for good oxygenation is satisfied by vigorous mixing of the seawater by waves breaking on the reefs and by currents flowing through and past the reefs. The photosynthesis of algae and sea grass also adds oxygen to the surface waters by day. Photosynthesis stops at night, so in sheltered places like lagoons, oxygen may be reduced at night by the respiration of all the animals and plants living there; however, water mixing continues on the seaward side of a reef, making this the ideal habitat for most corals. Sheltered areas close to the shore such as lagoons are less favorable not only because of potential oxygen reduction at night, but also because without free mixing with the open water, there may be excess heating from the sun during the day, or a build-up of brackish or polluted water flowing from the land.

# The Productivity of Coral Reefs

This subject is of great interest to ecologists because of the apparent paradox of such highly productive ecosystems, teeming with life, being situated in the relatively barren environment of tropical oceanic water. Measurements of primary production by coral reefs (the production of chemical energy through photosynthesis from a measured area of reef in a measured time) are generally at least an order of magnitude greater than in the surrounding ocean water (Hatcher 1997) and rival the highest levels recorded for natural systems. To be sustained, however, primary production requires more than simply the raw materials of photosynthesis (carbon dioxide, water, and sunlight): sufficient plant nutrients also need to be present in the environment for the growth of the photosynthetic organisms. The primary productivity then fuels the consumption and lives of all the organisms in the ecosystem— but, as noted above, tropical oceanic waters are naturally very low in nutrients. Indeed, corals grow best in low-nutrient conditions for reasons already explained. How, therefore, can reefs support the high production that is observed?

As is often the case, the answer is complex, consisting of several parts. First, a distinction must be drawn between gross production (the total amount of chemical energy produced within a certain space and time) and net production (the gross production minus the energy lost in respiration of all the different organisms within that same space and time). For an ecosystem to achieve surplus production, net production needs to be greater than zero; or, in other words, production to respiration ratios (P/R) should be greater than 1. Agricultural systems achieve considerable positive net production because they are subsidized by extra nutrients, and only half of the ecosystem is present—the consumers being elsewhere. Farmers can therefore deliver large crops to market. Natural systems are much less generous: over geological time the small annual global profit is now represented by such things as our accumulated fossil fuels—a small return when one considers the hundreds of millions of years involved. Returning to coral reefs, despite the practical difficulties of measuring production and respiration within a dynamic, submerged system, it has been shown that gross primary production over the whole reef is usually only slightly greater than total ecosystem respiration. A practical consequence is that the regular removal of large crops, for instance by fishing, is unlikely to be sustainable—as has been shown for many coral reefs.

Measurements of production and respiration on coral reefs show a wide range of values between different zones of a reef and between different reefs. However, the existence of high gross primary production, especially in the algal turfs of the shallow reef flat, has been confirmed, so the paradox remains: where do the nutrients come from?

Factors that add nutrients to a coral reef include the turbulent flow of water past the reef. Despite low concentrations of nutrients in the water, if organisms can take up and retain these nutrients as they flow past, they necessarily accumulate on the reef. It turns out that many reef-dwelling attached algae are able to take up nutrients from very low concentrations provided that the water is flowing; they grow well because water depleted by nutrient uptake is quickly replaced by less-depleted water. Consumption of plankton from ocean waters by reef organisms also plays a part in capturing nutrients on the reef. There is not much plankton in tropical ocean water, but the flow of water provides a continuous, dilute supply. In addition, oceanic zooplankton rises toward the surface at night, increasing the uptake by such animals as corals. A proportion of the nutrients captured by the corals in their plankton food is recycled to the zooxanthellae, directly sustaining their growth. A different but significant input of nutrients, especially nitrates, comes from the activities of nitrogen-fixing microorganisms such as Cyanobacteria, which are plentiful on reefs.

Finally, efficient recycling of nutrients within the reef ecosystem helps to retain nutrients on the reef. This efficiency is achieved by long food chains, including a wide variety of scavengers, detritus feeders, and microorganisms. All these various factors added together make it possible to resolve the apparent paradox.

## How Coral Reefs Form

The commonest coral reefs in the Caribbean are fringing reefs, so called because they occur as a fringe, a few tens or hundreds of meters from shore, running parallel to and more or less continuously along the coastline. This fringing ribbon of living corals is supported beneath by consolidated dead coral (or reef rock) and may be several tens of meters wide (from coastal edge to seaward edge) and a few tens of meters in vertical thickness. Fringing reefs generally rest on a rocky foundation, often an extension of the coastal rocks. These fringing reefs are a few thousand years old (relatively young in geological terms), an age that is recognized because the level of their foundations was above the level of the sea during the last ice age, which ended about 10,000 years ago. At the end of the ice age, the beginning of the present interglacial age, the melting ice caused sea levels worldwide to rise by about 100 m to reach roughly present-day sea level some 8000 years ago. Modern fringing reefs have grown since that time, often on reef deposits dating from an earlier interglacial age.

The size and shape of a fringing reef imply that reef growth is both upward toward the surface and outward from the land. The growth of corals, especially large so-called framework builders (e.g., *Montastraea* spp.), is the most important process, but wave action is also important since it breaks corals and piles up dead skeletons to form platforms for new coral growth. Additional smaller fragments, sand, and finer sediments combined with chemical and biological cementation processes help to fill in gaps and holes between and within the dead skeletons and fuse the whole mass together into reef rock. Growth of corals on reefs must, of course, exceed erosion in order to produce reef growth, and under normal conditions coral growth is sufficiently in excess of erosion to allow the whole reef to keep pace with a modest rate of sea level rise or a modest rate of coastal subsidence; thus living reefs remain just below sea level at low tide. Some corals can endure short exposure to air at low tide, but any drying by air and sun will kill them. Sinking sea level or rising coastlines therefore kill the corals by exposing them to the air, ultimately leading to raised fossil coral reefs, such as may be seen on many Caribbean islands.

Two other types of reef are known that have a longer history than fringing reefs: barrier reefs and atolls, known mainly from the Pacific, but representatives of both types occur in the Caribbean. Superficially these reefs resemble fringing reefs and are indeed just the same at the level of the reef-forming processes; however, they rest on fossil coral foundations that are much thicker than the shallow coastal platforms that normally underlie fringing reefs, and they are much further out to sea. Atolls—ring-shaped islands built on ring-shaped coral reefs surrounding large central lagoons—are found mainly well away from land in the Pacific and Indian oceans. Barrier reefs are massive systems of reefs and coral islands running parallel to continental margins but tens to hundreds of kilometers offshore. The living corals on atolls and barrier reefs have a very great thickness of reef rock beneath them, as demonstrated by drilling: the drill may pass through 1 km or more of reef rock before hitting bedrock. Reef-building corals grow only near the surface of the sea, so when the corals that formed the deepest reef rock were alive, they must have been near the surface, not 1000 m deep. This depth greatly exceeds the changes in sea level produced by the ice ages. The explanation (first suggested for atolls by Darwin) is that the bedrock beneath both barrier reefs and atolls has subsided during the growth of the reefs. In the case of atolls, these formed as fringing reefs on oceanic volcanoes that once reached the surface but subsequently subsided, leaving a massive ring of coral. Barrier reefs formed on slowly subsiding coastal platforms; the thickness of these reefs shows that their overall age is much greater than a few thousand years.

# Coral Reef Zones

Coral reefs are usually found some distance from the shoreline, with the area of calm water between shore and reef being called a lagoon. The width of lagoons can vary greatly, from a few meters to several kilometers. Lagoons are usually sandy, often with sea-grass meadows, and can be less than a meter to many meters deep. Some species of corals can tolerate the effects of limited water mixing and may be found in lagoons. These corals are often aggregated as small patch-reefs 1–5 m across, consisting of several species of coral and rising above the sandy bottom of the lagoon; however, the main reef is a much larger structure, within which at least four zones can be distinguished moving seaward from the lagoon (Goreau and Goreau, 1973). The widths of these zones depend on the width of the whole reef, which can vary greatly, from 10 m or less to 100 m or more.

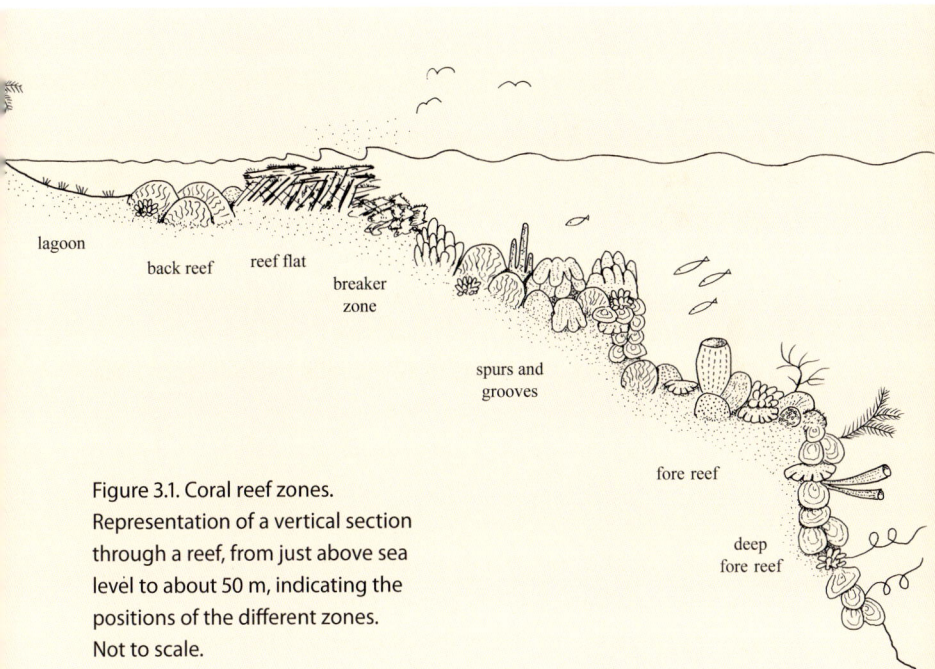

lagoon

back reef

reef flat

breaker zone

spurs and grooves

fore reef

deep fore reef

Figure 3.1. Coral reef zones. Representation of a vertical section through a reef, from just above sea level to about 50 m, indicating the positions of the different zones. Not to scale.

Plate 3.1. Viewed from the shoreline on the northeast coast of Trinidad, lagoon and reef flat at low tide.

The first zone, bordering the outer edge of the lagoon, can be referred to as the back reef or rear zone. Here the water is shallow and better mixed than in the lagoon; species such as brain corals are often found here. Seaward from the rear zone, the reef rises to a reef flat that may break the surface at low tide. The reef flat is washed by currents caused by waves breaking further out and may be dissected by channels through which the water washed in by the waves returns to the sea. The reef flat is usually composed mostly of dead coral skeletons overgrown with algae, colonial zo-anthids, and fire coral, with other coral species attached in crevices and to the sides of pools and channels. The front of the reef is referred to as the breaker zone or reef crest and is where the waves break. Before 1980, this zone all over the Caribbean was dominated by elkhorn coral; however, this species was very badly affected by white band disease (see chapter 4), and now in most places the breaker zone is composed largely of dead elkhorn coral skeletons overgrown by fire coral and encrusting coralline algae. Few other coral species can tolerate the high wave action within and just seaward of the breaker zone. Beyond the breaker zone, however, the reef slopes into deeper water and may then take the form of a series of spurs, with grooves between, leading into deeper water and toward the direction of approaching waves. The tops of the spurs may be 2–5 m below the water surface, whereas at the bottoms of the grooves, the depth may be 3–10 m. This spur and groove (or buttress and canyon) zone varies greatly depending on the underlying slope of the rock beneath the reef and also on the normal strength of the wave action. Generally, the more gentle the underlying slope and the greater the wave action, the longer the spurs. A wide variety of coral species inhabit this spur and groove zone.

Plate 3.2. View toward shore from between two coral spurs on the north coast of Jamaica.

Plate 3.3. View upward from the deep fore reef on the south coast of Curaçao; wire corals, feathery gorgonians, plate corals, and sponges are common.

These four zones—back reef, reef flat, breaker zone, and spur and groove zone—describe the structure of those parts of a reef that can be seen fairly easily from a boat and form most people's idea of a coral reef. Provided a suitable environment is available, however, the coral reef continues well beyond the spurs and grooves, often as lower-relief extensions of these features, into 10–25-m depth on what is usually called the fore reef. Broad, low-lying fingers of coral, with sand channels between, slope gradually downward, following the underlying rock. Coral species on the fore reef are generally similar to those found among the spurs and grooves. Along some coastlines, coral growth extends to 30–40 m and here, on the deep fore reef, coral species composition changes with the disappearance of shallow-water corals (e.g., species of *Acropora* and *Millepora*) and the addition of deeper-water species, especially large plate corals (e.g., species of *Agaricia* and *Mycetophyllia*).

Any particular coral reef will not necessarily resemble the description above in all its details. The shape and underlying slope of the coastline, the prevailing exposure to wave action, and the water quality all have very great effects on the development and species composition of reefs. For instance, Curaçao in the southern Caribbean has a steeply sloping and sheltered south coast with little development of reef zones; the rich and prolific coral community slopes steadily downward. In contrast, the north coast of Jamaica has cliffs and terraces and is exposed to moderate wave action from the trade winds; well-zoned reefs grow on the terraces and coral growth plunges over the cliffs. In both these cases, the coastal waters are very clear and clean, but close to the coasts of South and Central America where large rivers cause high turbidity, coral growth is restricted to a few tolerant species, and little reef development takes place. In these countries, good reef development is possible only offshore, such as along the Mesoamerican Barrier Reef system running 25 km or more offshore along the coasts of Mexico, Belize, Guatemala, and Honduras.

## Associated Animals

Coral reefs are sometimes called the "rainforests of the sea" on account of the three-dimensional nature of the reef environment, which provides numerous different habitats, and the very large number of different animals and plants living in these habitats. The number of species of noncoral animals and plants living on a coral reef runs into the thousands—far too many for a book of this size; therefore only those animals commonly seen directly on corals, or known to have an important role in coral ecology, will be described here, and in most cases only in general terms.

Plate 3.4. Cleaner goby resting on star coral, *Montastraea* sp.

Many animals live on, in, or attached to living corals. These include small fish that live on the surfaces of live corals, resting among the polyps. Large boulder or brain corals generally have at least one small resident fish—usually a goby 1–4 cm long with a conspicuous stripe along each side and belonging to the genus *Elacatinus* (of which there are many species). Most of these coral gobies are so-called cleaner fish that feed by removing (and eating) parasites, loose infected tissue, and other bits and pieces from the skins, mouths, and gills of larger fish. To receive this service, the larger fish come to coral heads where the cleaner fish live. The conspicuous stripes from nose to tail and certain behaviors of the cleaner fish are recognized by the larger fish and prevent the cleaner from being eaten by its customer (often a predator such as a snapper or grouper).

Other obvious organisms commonly seen on corals include worms that live in tubes embedded in the coral skeleton, extending their filter-feeding tentacles into the water above the coral tissue. The most conspicuous of these in the Caribbean is the Christmas tree worm *Spirobranchus giganteus*, so called because each of its pair of fine feathered tentacles forms a spiral cone, reminiscent of a tiny Christmas tree. The conical crowns of tentacles are about 1 cm high and occur in many different colors. There is no evident benefit to the coral from this association, but the worms are probably protected from predators by the coral's stinging cells and mucus. The worms themselves, and other animals living in close contact with coral tissues, may have some protective coating to avoid getting stung.

Peering more closely at the surface of living corals often reveals the presence of barnacles among the coral polyps. Coral barnacles can sometimes be very numerous, especially in turbid, well-fertilized environments where organic particles and plankton in the water provide plentiful food for these filter feeders.

Plate 3.5. Christmas tree worms on mustard hill coral, *Porites astreoides*.

Plate 3.6. Christmas tree worm in side view; the circular pink object is the operculum used to plug the tube when the worm withdraws.

As in the case of the filter-feeding worms, there appears to be no benefit to the coral; indeed, rather the opposite, since the barnacles and worms take up space and gather food the coral might otherwise catch. Coral barnacles belong to a specialized family of acorn barnacles, the Pyrgomatidae; they are conical shaped and sit directly on the surface of the coral.

Sponges are often in direct competition with corals for space on the reef, and a few species of sponge are able to bore into live corals, creating spaces for sponge tissue within the coral skeleton. Further growth of these boring sponges can ultimately kill entire coral colonies. Caribbean sponge species with this habit include *Cliona delitrix* and *Cliona tenuis*. The former is bright orange in color, sometimes with white spots caused by a white colonial

Plate 3.7. Barnacles on whitestar sheet coral, *Agaricia lamarcki*, showing coral tissues overgrowing the barnacles' shells.

Plate 3.8. Skeleton of sun-ray lettuce coral, *Leptoseris cucullata*, showing coral skeleton deposited on the barnacle shell.

Plate 3.9. Boring sponge, *Cliona delitrix*, erupting through the surface of massive starlet coral, *Siderastrea siderea*.

zoanthid living with the sponge. *C. delitrix* excavates large cavities within coral heads, and its vivid living surface is pockmarked with large oscula (exhalant pores for the sponge's filter-feeding system). *C. tenuis* is less conspicuous; semitransparent, colored gray brown with a feltlike texture, it excavates small chambers close to the surface of the skeleton, has numerous small oscula, and can take over and kill entire large coral heads (López-Victoria et al. 2006). Sponges are filter feeders, removing mainly very small organic particles such as bacteria and small phytoplankton. They flourish in well-fertilized coastal waters, and some experts have suggested that a high frequency of *C. delitrix* is an indicator of sewage pollution.

Several less easily visible but nevertheless important organisms bore into the bases of corals where the skeleton is laid bare by retreat of the tissues. These borers include algae, worms, sponges, barnacles, and bivalve mollusks. Bivalves bore into the coral skeleton by chemical means sometimes assisted by rasping with their shells. They excavate deep, flask-shaped burrows that, together with the borings of other organisms, weaken coral skeletons and make them more easily damaged by storms. Research comparing the number of bore holes in coral skeletons with the water quality where the corals were collected has shown a positive correlation between the number of borers and the fertility of the water. This relationship arises because all the boring animals feed, one way or another, on organic particles (plankton, bacteria, detritus) that in turn are plentiful in well-fertilized water. Thus

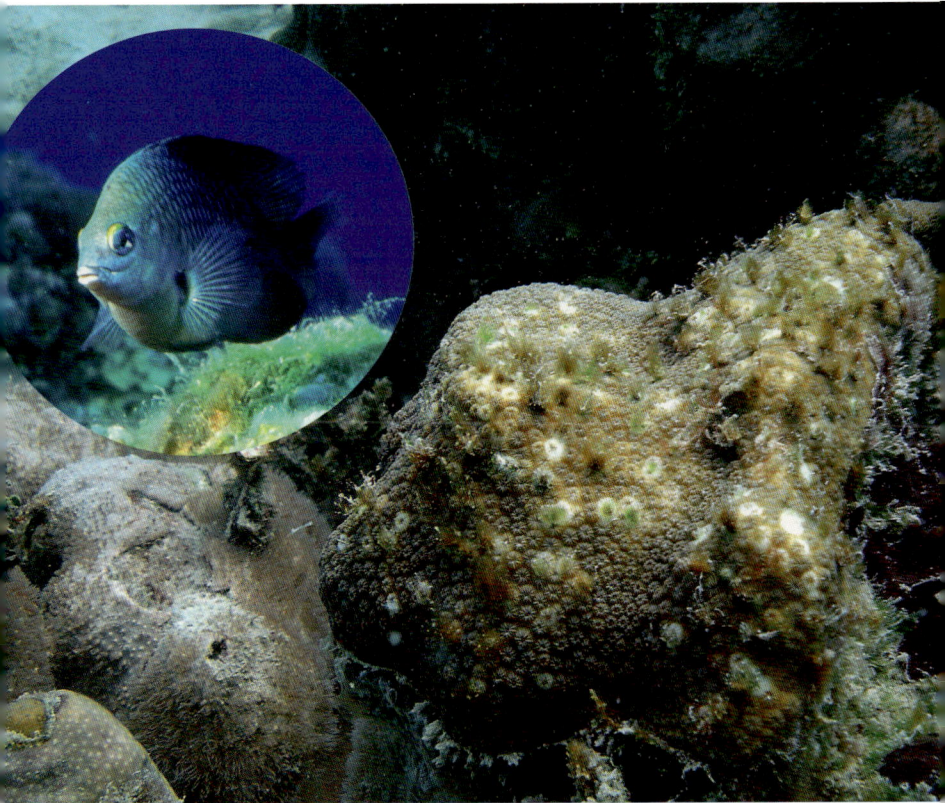

Plate 3.10. Lobed star coral, *Montastraea annularis*, damaged by the three spot damsel fish, *Stegastes planifrons* (*inset*). The adult fish are 8–10 cm long.

coral reefs growing in such locations suffer relatively high levels of erosion, especially during storms.

There are only a few specialized predators of coral, and a slightly larger number of occasional coral predators. The most conspicuous, but not necessarily the most important, are various fishes, including butterfly fish, trigger fish, puffer fish, and parrot fish. Some butterfly fish pick off individual coral polyps, but most have more varied diets that may not include corals as major components. Trigger fish and puffer fish have strong jaws and teeth and eat hard-shelled prey such as sea urchins and crustaceans, and occasionally they bite off pieces of coral. Parrot fish have jaws and teeth adapted for grazing algae from reef rock and dead coral; occasionally they may also graze living coral. Perhaps more important for coral mortality on Caribbean reefs, however, are territorial damsel fish, particularly the three-spot damsel fish, *Stegastes planifrons*, which do not eat coral but kill or damage corals growing in their territory, encouraging the growth of attached algae (algal gardens), which they do eat (plate 3.10). These damsel fish select structurally complex habitats to set up their territories, probably as protection from predatory fish. They used to live among staghorn coral thickets, but now that these have mostly disappeared, they select large colonies of *Montastraea annularis* and *M. faveolata* and have become a significant cause of partial mortality for these important corals (Garzón-Ferreira et al. 2005).

Invertebrate predators on corals include the fire worm (*Hermodice carunculata*), which feeds on branching corals, including fire coral and staghorn coral. This large worm (up to 20 cm in length) is well protected by bunches of sharp, detachable white bristles all along its body. The bristles are dramatically displayed if the worm is threatened, and if touched, the bristles can break off in the skin, causing a burning itch. Fire worms are often seen active on the reef, sometimes hunched over, with the tip of a coral branch thrust into the mouth. Less conspicuous predators include the coral snails *Coralliophila* spp.; these snails are small (often <1 cm long) and usually camouflaged by marine growths on the shell.

Plate 3.11. Among algae, fire worm, *Hermodice carunculata*, displaying its white irritant bristles.

They may be found lurking near areas of bare white skeleton (evidence of their recent foraging).

Important positive effects on corals are caused by grazing animals that eat algae. These animals clear space on which corals can settle and grow, so they indirectly increase coral abundance. Parrot fishes are important in this context, and so are sea urchins, especially the long-spined black sea urchin *Diadema antillarum*. More on these indirect interactions appears in the next chapter.

# 4 Current Status of Caribbean Coral Reefs

## Threats to Reefs

One might think that corals living in the pleasantly warm, clear waters of the Caribbean are in a stress-free environment; however, like all other animals, corals face numerous natural environmental challenges such as predation, disease, competition, severe storms, and changes in temperature, salinity, and turbidity of the water. These challenges place severe demands on their defensive capabilities. To survive and reproduce they must constantly maintain themselves by adequate feeding and growth. These natural requirements are sufficiently onerous that any long-term departure from the conditions under which the corals evolved is liable to cause extra stress, reducing the capacity of the corals to compete successfully and to thrive, or even to survive.

The current position is that since the late 1970s Caribbean corals have not been maintaining themselves successfully; there have been both gradual and rapid reductions in the abundance of corals on reefs throughout the Caribbean. Between roughly 1980 and 2000, average coral abundance on Caribbean reefs (measured as a percent of the surface area covered by living coral) declined from more than 50% to somewhere between 5 and 30% (Gardner et al. 2003). Not all reefs were affected to the same extent or in the same way, but virtually all countries and reef systems in the region suffered some decline in the abundance of corals. Since that time the general picture has been of lack of recovery, with a few reefs showing slight increases in coral abundance but most showing either stability at a new lower level or a continuing decline (Wilkinson 2008; Schutte et al. 2010). At present, Caribbean reefs with 40% or more live coral cover are rarities, with 10–20% being much more common. The general decline of coral abundance has been matched by reduction in the abundance of juvenile corals on reefs. For instance, in Curaçao, juvenile corals less than 4 cm in diameter have declined by nearly 55% over 30 years from an average of about 15 per square meter in 1975 to about 7 per square meter in 2005 (Vermeij et al. 2011). Reduced recruitment of course makes recovery more difficult. The space that used to be occupied by corals has mostly been taken over by attached algae of various species.

To determine the causes of the observed declines in coral abundance, we need to compare present conditions with those of the past. Unfortunately our knowledge of living coral ecology is fairly recent, and baseline data on original or pristine conditions are patchy at best. It has therefore often been difficult to identify precisely the conditions responsible for recent coral death; however, the evidence is accumulating that most of the apparently new observed environmental changes have been caused and continue to be caused, either directly or indirectly, by the activities of humans—locally, regionally, and globally (Mora 2008; Burke et al. 2011).

Some of the most important factors in coral death in the Caribbean include coral diseases, bleaching, increases in the sediment, nutrient, and pollutant content of runoff from the land, and hurricanes. Factors that lead indirectly to coral death include removal of organisms benefiting corals (e.g., some fishes), and changes benefiting the competitors of corals.

## Coral Diseases

Many apparently new diseases have emerged in the Caribbean over the last 30 years (Sutherland et al. 2004). The most serious historically has been white band disease (WBD), which appeared in the late 1970s and affects elkhorn and staghorn corals. These two species of *Acropora* used to dominate, respectively, the breaker zone and shallower parts of the fore reef all over the Caribbean; now they are relatively uncommon as live colonies, but their dead skeletons abound as rubble (Aronson and Precht 2001). In official response to this regionwide catastrophe, both Caribbean *Acropora* species were declared endangered in 2004 under the U.S. Endangered Species Act. WBD gets its name from its appearance: sharp bands of bare, white (dead) skeleton progress along the branches

Plate 4.1. Black band disease on elkhorn coral, *Acropora palmata*.

until the whole colony is dead; this can occur in a matter of days. Meanwhile, the trailing edge of the band becomes colonized by algae. WBD is now much less common and appears less virulent than in the past, partly because there is much less *Acropora* present, but also probably because the remaining colonies have some resistance to the disease (Vollmer and Kline 2008).

Several other "band" diseases are known, including black band disease (BBD, plate 4.1), affecting a wide range of species, yellow band (or blotch) disease (YBD), affecting several species but most prominently the star corals of the *Montastraea annularis* species complex, and white plague (WP), occurring in at least three types, distinguished by speed of progress and affecting a very wide range of species. The black band of BBD consists of a complex

Plate 4.2. Yellow band disease on star coral, *Montastraea faveolata*. The slow speed of advance is shown by the close approach of algal colonization to the edge of the pale diseased tissue.

Plate 4.3. White plague on massive starlet coral, *Siderastrea siderea*; the lack of algal growth close to the sharp edge indicates the rapid speed of advance of the disease.

of microorganisms, including Cyanobacteria and sulfur bacteria, that moves slowly across colonies, killing in weeks or months (Richardson et al. 2009). In YBD the pale yellow color indicates diseased and dying tissue; the disease moves very slowly and may not kill an entire colony in one episode, but it often results in large dead patches and fragmentation of the living colony, leading to partial mortality of 50% or more. YBD is an important factor in reef ecology, since the most commonly affected corals (*M. annularis* and *M. faveolata*) are major framework builders (Bruckner and Bruckner 2006).

The WPs are characterized by spreading areas of bare white skeleton with a sharp edge where the disease organisms are active; like WBD it can kill entire colonies quite rapidly. Dark spot disease is somewhat different; it is common on starlet corals (*Siderastrea* spp.), sometimes appearing to have little effect other than

Plate 4.4. Dark spot disease on massive starlet coral, *Siderastrea siderea*.

the dark spots themselves, but more usually infected colonies are pale and unhealthy looking and suffer gradual patchy mortality.

The causative agents of only a few of these diseases are known since samples of dying tissue contain many microorganisms and it is difficult to isolate and culture the particular one (or ones) causing the disease. The apparently new occurrence of many diseases in relatively few years implies a novel cause, and many suggestions have been made as to what this cause or causes may be, all linked in some way to modern human activities. One for which there is good evidence is that new stresses, for example, bleaching and pollution, have weakened corals and made them more susceptible to disease organisms. Bleaching incidents caused by increased temperature (see below) are often followed by an increase in disease (e.g., BBD, YBD), implying that this stress, or elevated temperature, contributes to infection and perhaps also to the virulence of the pathogen (Cróquer and Weil 2009). Another suggestion is that the disease organisms themselves are new to the Caribbean, having been introduced from elsewhere; possible agents of introduction include dust blown across the Atlantic from the Sahara Desert, floods from newly deforested land areas, and sewage. White pox disease (WPD), characterized by small irregular white patches and occurring on elkhorn coral, *Acropora palmata*, is apparently caused by a human enteric bacterium and therefore likely due directly to sewage pollution. WPD has been held responsible for a mass mortality of elkhorn coral in the Florida Keys National Marine Sanctuary during the 1990s. Another possible source of

Plate 4.5. Skeletal anomaly on boulder brain coral, *Colpophyllia natans*.

new diseases in the Caribbean is introduction from other oceans in the ballast tanks of commercial shipping. Shipping has certainly been responsible for many introductions of alien marine species all over the world, for example, western Atlantic comb jellyfish to the Black Sea, Australian barnacles to Europe, Pacific mussels to the Caribbean, and probably also the orange cup-coral *Tubastraea coccinea* from the Indo-Pacific to the Caribbean.

Skeletal anomalies, also referred to as neoplasms and tumors, have been observed on many coral species and may be classed as diseases. These usually take the form of enlarged and protruding skeletal features overlaid by paler tissues (plate 4.5). Some of these are caused by the coral's reaction to parasites or commensals, including fungal and algae infections, but for many there are no known causes.

## ■ Coral Bleaching and Global Warming

Technically speaking, coral bleaching is also a disease, but not an infectious one. Bleaching is a condition in which the zooxanthellae are expelled from the coral, leaving the living tissue transparent so the white skeleton shows through. There are different degrees of bleaching: corals may become pale or show pale or white patches, or go completely white. This condition is unlike that of WBD or WP in that the edges of the white patches are not sharp but grade gradually into adjacent colored tissues that still have zooxanthellae present. Further, algae do not grow on the white areas since these are still covered by living coral tissue. By expelling its zooxanthellae, however, a bleached coral loses an important part of its nutrition; bleaching therefore seriously weakens corals and may kill them. Bleached corals that recover by gradually replenishing their zooxanthellae nevertheless show reduced growth in that season and reduced reproduction the following year (Mendes and Woodley 2002).

Bleaching is not due to pathogenic organisms but is a reaction to unfavorable environmental conditions. It is caused mainly by

Plate 4.6. Bleached corals: (*left and above*) star coral, *Montastraea*; (*right*) finger coral, *Porites porites*.

unusually warm water, especially when combined with high light intensity such as occurs during calm weather. In the Caribbean, water temperatures near or exceeding 30 °C for more than a few days are liable to cause bleaching. Severe bleaching of many different species occurs if the damaging conditions persist for two or more weeks. Bleaching incidents are commonest between September and November, when water temperatures are near their seasonal peak, and recovery from serious bleaching may take several months. Thus following a bleaching incident in October, partially bleached corals may still be present as late as the following February. Unusually warm patches of water liable to cause coral bleaching can be detected by satellite, and early warning of the approach of such "hot spots" can be accessed over the Internet.

Some coral species show resistance to bleaching, and some individuals within a species appear to have more resistance than others, so during a bleaching incident the extent of bleaching of individual colonies varies greatly from fully bleached to apparently normal. At least part of this resistance seems to be due to the zooxanthellae, which are not all the same. Variations between types of zooxanthellae include different photosynthetic abilities at different depths (i.e., in different light intensities and wavelengths) and different functional efficiencies at different temperatures. Individual coral colonies may contain a mixture of two or more types of zooxanthellae, with proportions varying in different parts of the colony. The zooxanthellae ejected during bleaching are probably those for which the particular combination of temperature and light is so unsuitable that it damages the chemical reactions of photosynthesis, leading to a breakdown in the symbiotic relationship. The damage appears to involve production by the zooxanthellae of strongly oxidizing chemicals that are toxic to living tissues and cause oxidative stress (Weis 2008). The offending and damaged zooxanthellae are therefore ejected by the coral.

Bleaching incidents, hardly known in the 1960s and 1970s, appear to be increasing in both frequency and intensity, probably as a consequence of the present global warming trend. This trend is believed by most scientists to be caused by increasing levels of carbon dioxide ($CO_2$) in the atmosphere due mainly to the burning of fossil fuels to satisfy human demand for energy.

Other aspects of global warming with potentially negative effects on corals include sea level rise, caused by thermal expansion of water and extra meltwater from glaciers, and ocean acidification caused by increased amounts of dissolved $CO_2$ forming carbonic acid. The current rate of sea level rise is slow but continuous (about

2 mm per year) and unless widespread coral recovery and active coral growth resume, sea level may eventually rise well above reef crests, leading to increased coastal erosion. If the rate of melting of Arctic and Antarctic ice increases significantly, however, worst-case predictions are for one or more meters of sea level rise within a few years. The risk from ocean acidification stems from the fact that calcium carbonate, from which coral skeletons are built, is easily dissolved in acid. At present the oceans are slightly alkaline, but this is very slowly changing. While wholesale dissolution of reefs is unlikely in the short term, the secretion of skeleton during coral growth is sensitive to the acid/alkaline balance and may eventually become slower. This change in acidity could be very serious in view of all the other current stresses on corals (Hoegh-Guldberg et al. 2007; ISRS 2007). It is worth noting that some historical records of coral growth rate from several parts of the world show small but significant decreases over the past 30 years apparently correlated with rising seawater temperatures.

## Sediment and Pollution

Clearance or deforestation of coastal land for extraction of timber or for rural and urban development (agriculture, housing, roads, etc.) removes the natural vegetation and causes erosion of the topsoil. The runoff after heavy rain is brown with suspended sediment and runs into the sea, either directly or via a river, sometimes staining the sea for miles around. Following development, the runoff is often still brown and in addition to sediment may contain pollution arising from the various developments. This pollution might be sewage and wastewater from a town, village, or hotel; hydrocarbons and industrial additives from roads or factories; or agricultural chemicals and fertilizers from farmland. All these things are bad for corals (Fabricius 2005).

Suspended sediment makes the water turbid and may settle on corals, which either get buried or have to use energy to remove the sediment particles—energy that might otherwise have gone into

Plate 4.7. Artichoke coral, *Scolymia cubensis*, suffering from sedimentation: the outer margin is already dead (see arrow), and sediment bound in mucus remains on the disk.

growth and reproduction. Sediment settling on rocks deters coral larvae from settling and can bury them if they do. Coastal areas in the Caribbean are often heavily developed, and most of the human population lives in the coastal regions; thus many Caribbean coral reefs are chronically subjected to runoff containing sediment and pollution. This polluted runoff is difficult to prevent since tropical rainfall is often very heavy and can rapidly overload wastewater treatment facilities (if present).

Agricultural runoff and sewage constitute a particular problem since both contain fertilizers (plant nutrients). The presence of extra nutrients in the originally very clean seawater changes the ecology by directly favoring the growth of algae and indirectly favoring other competitors of corals (Szmant 2002). The growth of attached algae, when unconstrained by grazing herbivores, can smother low-growing corals and prevent coral planulae from settling, while extra planktonic algae make the water more nutritious for filter-feeding animals such as sponges. Attached algae and sponges both compete with corals for space on the reef. Most of the animals that bore into coral skeletons are filter feeders, so coral boring is also increased by nutrient pollution, which leads to weakening of coral skeletons and increases the damage and erosion caused by storms.

Plate 4.8. Green filamentous algae smothering finger coral, *Porites porites*.

Plate 4.9. Fish traps on a Jamaican beach.

## Fishing

Many Caribbean coral reefs experience very high fishing intensities. The results are the classic symptoms of overfishing: reduced populations of high-value fish such as snappers (Lutjanidae) and groupers (Serranidae) and reduced sizes of individual fish of all captured species. Caribbean fisheries on coral reefs are mostly multispecies fisheries, carried out by individuals or small groups in small boats powered either by oars or by outboard motors. Methods include fish traps, nets, hook and line, and spearfishing. The fish trap is a very common method, usually using quite small-mesh chicken wire over a large framework of wooden poles. Fish traps catch adults and juveniles alike, of all available species; almost all of the catch is consumed.

As far as corals are concerned, alterations in the populations of herbivorous and grazing fishes are important since these fishes eat the algae that compete for space with the corals (Hawkins and Roberts 2004). The main herbivorous coral reef fishes in the Caribbean are the parrot fishes (Scaridae) and the surgeon fishes (Acanthuridae). These two families often contribute the bulk of the catch from fish traps because they are normally very common on coral reefs. Although they are not generally regarded as high-value fish, they are consumed anyway. Reductions in the herbivore

Plate 4.10. View of an overfished reef in Jamaica; there are few corals, very few fish, and most of the space is occupied by algae.

populations and removal of most large individuals reduce the grazing on algae. The resulting unconstrained algal growth reduces the space available for coral recruitment and is particularly serious when, as is often the case, nutrient enrichment is simultaneously stimulating the growth of the algae.

## ◼ Sea Urchins

A final twist of the herbivore tale, which has little to do with fishing and more to do with new diseases, was the mass mortality experienced in 1983–84 by the Caribbean long-spined black sea urchin *Diadema antillarum* (Lessios et al. 1984). Before the mass mortality, this animal was very common on shallow coral reefs throughout the Caribbean. It grazes algae from reef rock and is thought to have been very important in controlling the abundance of algae on coral reefs prior to 1983. The mass mortality was swift: the infective agent swept from the region of Panama around the Caribbean with the prevailing currents, killing about 95% of all *Diadema*. Following the mortality, a large increase in algal growth was observed on coral reefs. There is some disagreement over whether the algae directly smothered and killed corals or simply grew on corals killed by other factors such as disease (WBD was active at this time) or hurricanes; however, algae certainly hindered any coral recovery that might otherwise have occurred, and the increased proportion of algae has generally been maintained ever since. Recovery of populations of *Diadema* has been slow and localized. The infective agent that caused the mass mortality has not been identified.

Where *Diadema* populations have recovered, there is evidence both of reduced algal abundance and of increased coral settlement and growth, suggesting that this mass mortality probably did contribute to the decline in corals between 1980 and 1990, even if only by preventing recovery from other impacts.

Plate 4.11. Long-spined black sea urchins, *Diadema antillarum*, important grazers of algae on coral reefs.

Plate 4.12. Hurricane damage to a dead reef of elkhorn coral, *Acropora palmata*, probably originally killed by White Band Disease.

## Hurricanes

Clear damage to corals is caused by hurricanes, which smash them with waves, roll them around, and pile them up in heaps. This action is a natural part of reef building, but in the context of all the other modern threats facing corals, it is simply an additional stress, slowing recovery from other simultaneous stresses and reducing the chances of complete recovery. There is also the possibility that global warming will lead to more frequent and more intense hurricanes. Indeed, global warming with its increased risks of bleaching, ocean acidification, sea level rise, and storms is probably the most serious threat to coral reefs worldwide.

Plate 4.13. Colony of massive starlet coral, *Siderastrea siderea*, overturned by hurricane waves during Hurricane Ivan, 2004, Jamaica.

# Coral Reef Monitoring

It is possible to tell when something is changing only if you measure it repeatedly. Coral reef monitoring has grown from a rare activity in the 1970s, when coral reefs were thought to be stable ecosystems, to the present, when several monitoring schemes are being actively pursued locally and internationally by both scientists and concerned amateurs. Recognition of the deteriorating condition of coral reefs came initially from observations of large changes over short timescales, in particular, hurricane damage in the Caribbean and crown-of-thorns starfish damage in the Indo-Pacific. Studies of coral abundance "before" and "after" became necessary to measure the extent of damage, and these developed into regular monitoring as the extent of the deterioration of reefs became clear. In the Caribbean, such monitoring schemes include Caribbean Coastal Marine Productivity (CARICOMP), Atlantic and Gulf Rapid Reef Assessment (AGRRA), Reef Check, and the Global Coral Reef Monitoring Network (GCRMN), all of which publish periodic reports (e.g., Lang 2003; Marks 2007; Wilkinson 2008) that are made available through dedicated Web sites.

Most monitoring methods include estimates of live coral cover (percent of available space occupied by living corals), and this statistic is widely accepted as a rough guide to the health of a coral reef. The estimate is usually made by placing a marked line, known as a transect line, on a reef and counting and measuring live corals along the line; this procedure is repeated several times to obtain an average for the whole reef. Lines should be placed at random to avoid bias. There has been much discussion about the number and length of lines required to obtain a reliable estimate, and about the best way of counting the corals. Statistically, the more lines the better, and a generally accepted (and convenient) length of line is 10 m; however, an important constraint is the limited time that divers can spend underwater, so while 20 or 30 lines might be ideal, in practice between 5 and 10 lines per reef is more common. Overlapping the time constraint is the counting method and the skill

Plate 4.14. Coral reef monitoring by the video-transect method, using a tape measure as a marked line.

of the divers: the more information required (e.g., species of coral, size of colonies, other non-coral organisms, precision of measurement), the more skill is needed in both diving and reef ecology, and the longer it takes. A common and simple method for corals and other attached organisms is to mark a series of points along the line and record what is under each point: a 10-m line with 100 evenly spaced points gives percent cover directly, since if 30 of the points lie above living coral, the coral percent cover is 30%. Data can be recorded with a pencil on a white plastic slate roughened with sandpaper. A different data recording method that saves time underwater is to record the whole line using an underwater video camera, providing a direct record that can be examined at leisure later by more than one person. A disadvantage is that it is more difficult to identify corals from video than from the living specimens.

## Conservation

The goal of coral reef conservation is to maintain or increase the percent cover of live coral and the abundance of any other organisms that seem to have declined or are thought to have special value. These other organisms, such as certain fishes, must also be monitored to establish a baseline and to detect whether conservation has been successful.

Most coral reef conservation takes the form of protecting, or attempting to protect, a stretch of reef from the various threats described above, in the hope that natural processes of settlement and recruitment of coral reef species can be relied on to repair any damage that has already been done. This approach, of course, assumes adequate reproduction of surviving populations of coral reef species to supply the required larvae. In practice, coral reef conservation involves managing the people who live and work near or on the reef to reduce any activities that may be causing damage. Areas of protected reef are sited in Marine Protected Areas (MPAs) and may be declared by government and underpinned by legislation, or they may be reefs protected by voluntary agreements of stakeholders (often a first step in obtaining official recognition and legislative protection). The level of protection varies greatly, from virtually no protection (paper parks) to strict rules enforced by a team of wardens. Management and enforcement are costly and therefore often inadequate in countries with relatively weak economies. Ideally, protected reefs should be no-take zones from which the removal of anything is legally and actively prevented (experiments have repeatedly shown that no-take zones greatly increase populations of fishes and also may improve the

Plate 4.15. Poster aimed at tourists at a Jamaican Marine Protected Area (MPA).

fisheries outside the protected area). Reefs should also be protected from damaging runoff from the land. This can be achieved at least in part by preserving or replanting the natural vegetation on the shore (e.g., mangroves), by soil conservation inland, and by adequate wastewater treatment in nearby towns. Unfortunately it is rarely possible to achieve this ideal state, and it has been suggested that 90% of Caribbean reefs are at risk from the combined threats described above, with the risk being high or very high for 55% of reefs (Burke et al. 2011).

In practice it is in any case possible for local conservation managers only to protect against the threats that arise locally. These include local extractive threats (e.g., fishing, shell and coral collection), direct damage (boats, anchors, SCUBA divers, etc.), pollution threats (runoff and wastewater with suspended sediment, sewage, nutrients and other pollutants), and commercial developments nearby. Local managers cannot protect against new diseases, bleaching, or hurricanes, nor act by themselves to mitigate global warming! Much of the work of MPA managers and staff involves minimizing direct damage to corals from anchors by offering advice and providing mooring buoys, regulating potentially dangerous activities such as spearfishing and speedboats, and attempting to influence the public as well as local and national governments in favor of fisheries management, conservation of soil, and reduction of pollution.

A different type of conservation is one that attempts to undo any damage directly, by constructing new reefs, or growing new corals on damaged reefs. This is the area of artificial reefs and coral gardening. Artificial reefs can be made from a wide variety of materials and structures, ranging from natural boulders through prefabricated shaped blocks to large waste-steel structures such as old ships and rail cars. All these objects when placed on the seabed tend to attract shoals of fish and to be settled by sessile organisms, including algae and corals. Their success as artificial reefs can be judged by the variety of organisms attracted to them, and by their longevity. Virtually all are somewhat successful in the short term, unless they turn out to be toxic or pose an unexpected hazard to shipping. In the longer term their components may be dispersed if not heavy enough (e.g., vehicle tires) or they collapse from corrosion (steel) or become buried in sediment. In the meantime, however, they may provide an interesting feature for divers or fishers as an alternative to a natural coral reef. Similarly, reconstruction of a natural coral reef damaged by, for instance, ship-grounding, can also be carried out using concrete and cement and funding from the ship's owners or insurers.

Cultivating corals to populate a reconstructed or artificial reef, or to repopulate a degraded reef, can be achieved by breaking healthy coral colonies into small pieces, growing them to a larger size in a coral nursery, and then planting them out on the reef using special adhesives. More than 100 pieces, each with several live polyps, can be produced from a single healthy colony of sufficient size, so production can be rapid provided there is healthy growth of the pieces. The problem of reduced genetic diversity—the coral pieces from a single colony are all genetically identical—can be avoided by using several or many different parent colonies for each species in culture. Repopulation has been carried out successfully in projects in many parts of the World. Its utility as a general conservation technique, however, is limited by very high labor costs when applied to the huge areas of reef that need repopulating with corals, and particularly when the conditions still exist which led to reef degradation in the first place. The technique might therefore be suitable only as a backup to standard conservation methods for limited areas and particular locations.

It tends to be assumed by most environmentally conscious people that coral reefs are very precious and obviously worth conserving; however, not all people would set the protection of coral reefs above the need for economic development. It has therefore become increasingly necessary to provide economic reasons (as opposed to ethical or aesthetic reasons, although these can also be given a monetary value) for conserving coral reefs. Luckily this is easy to do since coral reefs provide several clear economic benefits. First, they protect the coastline during storms; this value can be equated with the cost of rebuilding existing coastal property or of building a defensive barrier instead of the reef. Second, if managed correctly they can provide a sustainable harvest of fish and other organisms, the value being the market value of the catch. Third, they attract tourists who come to admire the colorful fish and corals, and through natural erosion processes coral reefs provide the sand for the tourists to relax on; tourism value can be calculated from the amount of money spent on holidays. These types of valuation, particularly along coastlines with existing tourism developments, can mount up to many millions of U.S. dollars per kilometer of reef.

In addition to the economic arguments, international pressure is applied to the governments of Caribbean countries to sign on to various international environmental treaties and agreements, such as the Ramsar Convention, encouraging the protection of wetlands including coral reefs, and the Convention on International Trade in Endangered Species (CITES), which regulates international trade in a huge range of animals and plants, including corals. In addition, there is the United Nations Environmental Program's Cartagena Convention, applying to the wider Caribbean and comprising, in turn, the Oil Spills Protocol, the Specially Protected Areas and Wildlife in the Wider Caribbean (SPAW) protocol, and the Pollution from Land Based Sources and Activities (LBS) protocol. These conventions oblige signatories to put in place the means to apply the terms of the agreements, providing an excellent starting point for government support of conservation. Together with economic valuations of coral reefs, signing the relevant international conventions should make the argument for conservation compelling for governments, but generally does not. The results more often are strong-sounding but underfunded policies that can be circumvented when required.

Factors that tend to prevent adequate coral reef conservation in the Caribbean center around the developing economic state of many of the countries involved. In these countries the common problems of high unemployment, low wages, poverty, constrained educational provision, and high population growth affect all attempts to protect natural resources, including coral reefs. Overall the political climate is geared to encouraging commercial developments, and where these developments run counter to the needs of conservation, it is usually the development that wins. Similarly, high unemployment and poverty make the control of fishing difficult, both politically and in practice. The same pressures on poor people lead to unplanned coastal communities (squatter communities) that lack the infrastructure to conserve soil or to dispose of waste. Farming practices do not in general include soil conservation or organic farming, except through necessity when fertilizers and pesticides run out.

Thus although the governments of developing Caribbean nations have declared many marine parks and MPAs, very few of these contain no-take zones. MPA managers usually have access to plans for new developments nearby, through the introduction in many countries of compulsory Environmental Impact Assessments (EIAs), and there is generally some public discussion before planning consent is granted; however, the advice of MPA managers is frequently set aside.

Conservation efforts are either supported by government, often with external funding, or run by nongovernmental organizations (NGOs) that struggle both for local recognition and for funding. In both cases the funds consist mainly of grants from the overseas aid budgets of developed countries, and from international donors with environmental interests such as the Inter-American Development Bank, the Global Environment Fund, the World Bank, and UNEP; however, most funding is earmarked for particular projects and has a particular and relatively short-term duration. Lack of consistent funding therefore often hampers the long-term efforts required for coral reef conservation.

Funding of MPAs is better in Caribbean countries with stronger economies where long-term funding to support MPA management and staff comes directly from government, and short-term grants are needed only for special projects. In consequence of the stronger economies, there is less poverty and better educational provision, leading both directly and indirectly to better fisheries management, fewer unplanned settlements, and better provision for wastewater treatment.

In summary, many committed people are working hard at coral reef conservation in the Caribbean region, and recognition of the value of coral reefs is increasing. It remains to be seen whether this is enough to reverse the declines seen since the 1970s, especially in the face of continued global warming and ocean acidification.

## Current Trends and Predictions

Lack of recovery of coral abundance on Caribbean reefs, with some variation in both directions, seems to be the general trend over the years since 2000. The *Acropora* species in particular still show little recovery, although the live colonies that remain appear to have increased resistance to WBD. The *Montastraea* species, major reef framework builders, continue to suffer from YBD, which appears to be increasing in virulence as elevated temperatures and bleaching incidents become more frequent. Most of the other threats to corals, described above, continue to affect reefs throughout the region, and global warming seems likely to continue for many decades into the future.

Conservation action will probably increase, particularly in countries with stronger economies, as more people recognize the serious consequences of vanishing reefs. Fisheries management should be the easiest measure to implement, with no-take MPAs and regulations such as the banning of small mesh sizes on fish traps. Such regulations for small island developing states at present entail major political and enforcement challenges, and implementation will therefore be gradual. Control of polluted runoff from land is more difficult since the types and sources of pollution are so many and varied; however, mitigation measures are certainly possible and can be expected to increase. Introduction of alien organisms by commercial shipping can be lessened by new international legislation (for instance on sterilization of ballast water), but although such legislation will probably be introduced, commercial shipping itself will probably continue to increase. And global warming, with increased bleaching, sea level rise, acidification, and possibly increased storm action, will also continue.

The reaction of coral communities to this mixture of increased conservation and simultaneous increases in some stresses will be reflected in changes in species composition and in adaptation by some species. Species that show tolerance to turbidity, or to warm water, will increase proportionately on reefs affected by these stresses, whereas more sensitive species will diminish and may disappear. Similarly, adaptations that increase tolerance of the changes may evolve. Corals have a long fossil history, and major changes in the past have not eliminated them, so it can be predicted that coral, as a type of animal, will survive. The diverse coral communities that have successfully built reefs over the last few million years may not survive intact, however, and those species that remain may not build reefs. Many of the local and global stresses have the combined potential to both reduce coral growth and reproduction (by bleaching and acidification) and promote bioerosion (by adding to the fertility of the water). Thus the balance between overall reef growth (net calcium carbonate accretion) and overall erosion may soon become negative, and coral reefs may disappear from many tropical coastlines.

Current recommendations by concerned scientists are for rapid local action to eliminate overfishing and pollution and international action to reduce the excessive output of carbon dioxide (Aronson and Precht 2006; Hoegh-Guldberg et al. 2007). Rapid and effective local action is seen as the best hope for giving coral reefs a chance of surviving the increasing stresses of global warming.

# 5 | Caribbean Fossil Corals

## A Brief History of Reefs

One may hear it said that corals, with their very long fossil history, are among the world's most ancient creatures. The impression given is that coral reefs have flourished in tropical seas for hundreds of millions of years. While true in some respects, this is by no means a full picture (Veron 2000, 2008). Reefs produced by biological processes (biogenic reefs) have indeed been present for hundreds of millions of years, but not necessarily as *coral* reefs. The earliest biogenic reefs appeared in the Precambrian age some two to three thousand million years ago, when the only life forms present were bacteria and single-celled organisms. They were built by Cyanobacteria, which formed mats on the shallow sea bed and were capable of photosynthesis, producing oxygen, and with precipitation of calcium carbonate as a byproduct. The mats trapped the precipitate and other sediments to form mounded structures called stromatolites, consisting of compacted calcium carbonate and silicate sediments. When numerous, these stromatolites formed reeflike habitats. Stromatolite reefs still exist today in a few places, including locations in the Bahamas.

Multicellular organisms appeared during or shortly before the Cambrian age 550–500 million years ago, and archaeocyathids (probably calcified sponges) were the first to join stromatolites to form reefs and reef mounds. Over the next three ages (Ordovician, Silurian, and Devonian, 500–370 million years ago), several

Plate 5.1. Raised reef forming a cliff close to the north coast of Curaçao.

other multicellular groups of organisms with calcium carbonate shells or skeletons evolved and became reef builders. These included stromatoporoids (calcified sponges), calcareous algae, and primitive tabulate and rugose corals (skeletons made from calcite, a different form of calcium carbonate from the aragonite of scleractinians). Sediments produced by a variety of other calcareous shelled organisms (bryozoans, brachiopods, mollusks, etc.) added bulk to these reefs by binding and trapping. The Devonian (sometimes known as the age of the fishes) ended with a mass extinction (Veron 2008), and little reef development occurred over the next 60 million years of the Carboniferous age (the age of coal). Stromatolite and algal reef mounds appeared again in the Permian age (300–250 million years ago), but another mass extinction at the Permian-Triassic boundary eliminated reef development for several more million years.

By the middle Triassic (about 230 million years ago), scleractinian corals had made their first appearance (Stanley 2003). These were early forms of modern corals and probably contained zooxanthellae. Scleractinian corals diversified further as the age progressed (despite a very severe mass extinction event at the end of the Triassic), and in the Jurassic age (200–150 million years ago), while dinosaurs were evolving on land, the first true coral reefs appeared. Many reefs were not dominated by corals, however, which often shared space with stromatolites, stromatoporoids, and rudist bivalves. Rudists were large bivalve mollusks with heavy shells that became dominant reef builders in the Cretaceous (145–65 million years ago, the age of the dinosaurs), while corals were less important; however, the extinction event at the end of the Cretaceous that finished off the dinosaurs also saw the end of the rudists. The Cenozoic age, comprising the past 65 million years, saw the greatest development of modern coral reefs, especially during the last 25 million years as the earth cooled toward the ice ages.

## The History of Caribbean Coral Reefs

The earth's climate cooled gradually during the Cenozoic, with the most severe ice ages occurring during the last 2 million years, largely in the Pleistocene period. The ice ages comprised a series of glaciations and shorter, warmer interglacials. Coral reefs existed throughout this period, but sea level fluctuated greatly (as much as 140 m lower during glaciations); fossil Pleistocene coral reefs are therefore spread over a range of depths, many occurring above present sea level. These raised reefs may have formed at higher sea levels during interglacials, or (more often) may have resulted from geological forces raising the bedrock. Such reefs and reef deposits can be found on or near the coasts of many Caribbean islands and are mostly a few tens of thousands to hundreds of thousands of years old. Some Caribbean islands show two or more raised reefs parallel to the shore, and of increasing height and age. In the most recent fossil reefs, the preservation of corals is so good that modern species can be clearly recognized. Older fossil reefs, usually found further inland, often contain corals that are now extinct in the Caribbean.

Investigation of the origins of present-day Caribbean corals requires an understanding of pre–ice age changes, especially the relative movements of the tectonic plates making up the earth's crust. The Caribbean lies on a relatively small plate called the Caribbean

Plate 5.2. Raised reef close to modern sea level on the south coast of Jamaica, showing brain corals, *Diploria strigosa*.

Plate 5.3. Eroded fossil star coral on a beach in Jamaica, showing corallites.

Plate, which is moving slowly east in relation to its neighboring tectonic plates. In the distant past it was actually in the Pacific! Back some 60 million years ago, the Atlantic Ocean was narrower, North and South America were not joined through Panama, and an ancient warm sea called the Tethys, running through what is now the Middle East and the Himalayas, joined the Atlantic to the Western Indo-Pacific. There was therefore a free flow of warm surface water from east to west around the entire globe. Over the succeeding millions of years, Africa and particularly India moved north, narrowing the Tethys Sea, while the Atlantic gradually widened. The sea passage through the Tethys closed about 15 million years ago; the Mediterranean Sea remains at what was its Atlantic end, but is no longer warm enough to support coral reefs. Thus the Caribbean region became more isolated and was finally cut off

Plate 5.4. Fossil reef deposits in Curaçao, showing branches of elkhorn coral, *Acropora palmata*.

from the eastern Pacific by the Isthmus of Panama about 3 million years ago (O'Dea et al. 2007).

The step-by-step changes in ocean circulation due to the gradual closing of these sea passages, combined with gradual cooling and finally with ice-age climatic variations, caused severe ecological changes in the Caribbean. Many species and genera of corals became extinct in the Caribbean, although some that used to live there still exist in the Pacific (e.g., *Pocillopora, Stylophora*). The modern Caribbean coral fauna is therefore a small, isolated remnant or relic of earlier times, derived largely from the coral fauna of the Tethys (known from fossil evidence), somewhat genetically different from the present Indo-Pacific coral fauna (Fukami et al. 2004b) and with a much lower diversity. Luckily, however, this remnant includes a few large reef framework building species such as elkhorn coral *Acropora palmata* and star corals of the *Montastraea annularis* complex that have helped build the impressive modern reefs seen today in the Caribbean region.

# Glossary

**azooxanthellate.** Lacking zooxanthellae.

**bioerosion.** Erosion caused by the activities of living organisms, as in rasping, excavating, and burrowing.

**brooding.** Retaining eggs for internal fertilization and development of larvae.

**cilia.** Microscopic hairlike organelles projecting outside cells and able to beat to propel the bearer through the water or to move particles across the living surface (singular: **cilium**).

**class.** Major division of a phylum.

**cnidae.** Microscopic organelles within stinging cells of a cnidarian that eject a stinging or entangling thread when stimulated (singular: **cnida**).

**Cnidaria.** Phylum of simple, radially symmetrical animals in either polyp or medusoid form, such as hydroids, jellyfish, sea anemones, and corals.

**coenosteum.** Porous skeletal material between corallites.

**colonial.** State in which many individual polyps are joined to others in the same colony and belong to the same clone.

**columella.** Central skeletal structure in a corallite.

**corallite.** The skeleton of a coral polyp.

**costae.** Name given to those parts of septa that extend beyond the corallite wall (singular: **costa**).

**cup-coral.** Coral in which each polyp is more or less isolated within a simple cup-shaped corallite, often solitary but may be joined to others.

**Cyanobacteria.** Phylum of photosynthetic bacteria, previously called blue-green algae.

**ecosystem.** Diverse group of organisms within their recognizable and characteristic environment, operating as an interdependent system.

**family.** Group of related genera.

**gametes.** Eggs and sperm.

**genus.** Group of closely related species (plural: **genera**).

**glacial age.** An ice age, during which time sea levels fall because a large amount of water is frozen as thick ice sheets on land.

**goby.** Member of the family Gobiidae, a group of small mainly marine fish that live on the seabed.

**gonochoric.** Having separate male and female individuals or colonies.

**gorgonian.** Member of the order Gorgonacea, a group of colonial cnidarians with flexible skeletons forming branched, plantlike growths on reefs.

**Hermaphrodite.** Individual or colony producing both eggs and sperm.

**Hydrozoa.** Class of cnidarians usually with both polyp and medusa generations in their life cycles and with small simple body structures.

**interglacial age.** Period of time between ice ages, during which sea levels rise because the extensive ice sheets on land melt.

**meandroid.** Coral structure in which polyps and corallites are not separate but merge with one another to form meandering rows within grooves containing several separate mouths.

**meiosis.** Form of cell division in which cells with two sets of chromosomes (diploid) divide twice to produce gametes with one set of chromosomes (haploid).

**mesenteric filament.** Thread bearing many cnidae within the digestive cavity of corals and sea anemones which are used in digestion but can also be extruded for aggressive or defensive purposes.

**monophyletic.** Group that has evolved from a single ancestral species.

**morphology.** The form of the body of an animal or plant; study of such forms.

**nitrogen-fixing.** Able to incorporate atmospheric nitrogen into chemical compounds that can be used for nutrition by organisms.

**nutrients.** Chemical compounds essential for life, especially inorganic molecules, such as nitrates, phosphates, required by plants.

**order.** Major division of a class—for example, order Scleractinia.

**organelle.** Subcellular structure contained within cells.

**pali.** Projections arising from the inner margins of fused septa (singular: **palus**).

**paliform lobe.** Projection arising from the inner margin of a septum.

**photosynthesis.** Biochemical process by which plants use the energy of sunlight to produce sugar and oxygen by combining water and carbon dioxide.

**phylum.** Major division of a kingdom—for example, Cnidaria in the animal kingdom (plural: **phyla**).

**plankton.** Community of small organisms drifting in open water.

**planula.** Motile cnidarian larval form consisting of a more or less elongated ball of ciliated cells, its function being to locate a settlement site and to metamorphose into a polyp.

**polyp.** One of the two main cnidarian body forms, consisting of a cylindrical, radially symmetrical column, with one end fixed to the substrate and a mouth surrounded by tentacles at the other end.

**polyphyletic.** Describing a group that contains two or more different ancestral lines.

**primary septa, secondary, etc.** Types of septa, judged by size and by repeating radial pattern, primary being largest, secondary being smaller, tertiary being smaller still, and so on.

**production.** Amount of energy or biomass (depending on unit of measurement) produced by a stated quantity of organisms over a stated time period, primary production being that produced by plants, secondary production, by herbivores, and so on.

**recruit.** Juvenile organism regarded as having joined a population of its species.

**reef.** Outcrop of hard substrate on the seabed, especially one created by growth of corals.

**reef rock.** Type of sedimentary rock formed from consolidated reef (especially coral reef) deposits.

**respiration.** Biochemical reactions by which organisms obtain energy to carry out their life processes, typically releasing energy by combining sugars with oxygen to form water and carbon dioxide.

**sea anemone.** Usually solitary cnidarian polyp, lacking a solid skeleton, belonging to the order Actiniaria, class Anthozoa.

**Scleractinia.** Order within the class Anthozoa to which most hard corals belong.

**septa.** Skeletal plates radially arranged within a corallite, dividing the digestive cavity into numerous pouches (singular: **septum**).

**septo-costa.** Name describing the entire structure of a septum that extends as a costa beyond the corallite wall (plural: **septo-costae**).

**spawning.** Releasing eggs and sperm into the water for external fertilization.

**species.** Organisms belonging to a single gene pool, that is, very closely related organisms that interbreed freely with each other; abbreviated **sp.** (singular) or **spp.** (plural) following a genus name.

**symbiont.** Organism living in close association with a different species, especially where there are benefits to both partners in the **symbiosis**.

**syn.** Abbreviation for synonym, an alternative name used for a species.

**taxonomist.** One who practices taxonomy, that is, names species and arranges them in groups according to their evolutionary relatedness.

**tectonic plate.** Large section of the earth's crust that moves very slowly in relation to neighboring plates.

**transect line.** Line traversing one or more habitats along which measurements are taken.

**turbid.** Description for water in which the presence of many fine particles reduces the transmission of light, rendering the water unclear.

**zoanthids.** Mainly colonial cnidarians that resemble sea anemones, belonging to the order Zoantharia, class Anthozoa.

**zooxanthellae.** Microscopic single-celled dinoflagellate algae that live inside the cells of certain animals, particularly corals, in a symbiotic relationship—hence, **zooxanthellate**, having zooxanthellae.

# References

Aronson, R. B., Precht, W. F. 2001. White-band disease and the changing face of Caribbean coral reefs. *Hydrobiologia* 460: 25–38.

Aronson, R. B., Precht, W. F. 2006. Conservation, precaution, and Caribbean reefs. *Coral Reefs* 25(3): 441–50.

Birkeland, C., editor. 1997. *Life and death of coral reefs*. New York: Chapman & Hall. 536 p.

Bruckner, A. W., Bruckner, R. J. 2006. The recent decline of *Montastraea annularis* (complex) coral populations in western Curaçao: a cause for concern? *Revista de Biología Tropical* 54 (supplement 3): 45–58.

Burke, L., Reytar, K., Spalding, M., Perry, A., editors. 2011. *Reefs at risk revisited*. Washington (DC): World Resources Institute. 112 p.

Cairns, S. D. 1982. Stony corals (Cnidaria: Hydrozoa, Scleractinia) of Carrie Bow Cay, Belize. *Smithsonian Contributions to the Marine Sciences* 12: 271–302.

Cróquer, A., Weil, E. 2009. Changes in Caribbean coral disease prevalence after the 2005 bleaching event. *Diseases of Aquatic Organisms* 87(1–2): 33–43.

Diekmann, O. E., Bak, R.P.M., Stam, W. T., Olsen, J. L. 2001. Molecular genetic evidence for probable reticulate speciation in the coral genus *Madracis* from a Caribbean fringing reef slope. *Marine Biology* 139: 221–33.

Fabricius, K. E. 2005. Effects of terrestrial runoff on the ecology of corals and coral reefs: review and synthesis. *Marine Pollution Bulletin* 50: 125–46.

Fenner, D. P. 1993. Species distinctions among several Caribbean stony corals. *Bulletin of Marine Science* 53(3): 1099–116.

Fenner, D. 1999. New observations on the stony coral (Scleractinia, Milleporidae, and Stylasteridae) species of Belize (Central America) and Cozumel (Mexico). *Bulletin of Marine Science* 64: 143–54.

Fukami, H., Budd, A. F., Levitan, D. R., Jara, J., Kersanach, R., Knowlton, N.

2004a. Geographic differences in species boundaries among members of the *Montastraea annularis* complex based on molecular and morphological markers. *Evolution* 58: 324–37.

Fukami, H., Budd, A. F., Paulay, G., Solé-Cava, A., Chen, C. A., Iwao, K., Knowlton, N. 2004b. Conventional taxonomy obscures deep divergence between Pacific and Atlantic corals. *Nature* 427: 832–35.

Fukami, H., Chen, C. A., Budd, A. F., Collins, A., Wallace, C., Chuang, Y.-Y., Chen, C., Dai, C-F., Iwao, K., Sheppard, C., Knowlton, N. 2008. Mitochondrial and nuclear genes suggest that stony corals are monophyletic but most families of stony corals are not (Order Scleractinia, Class Anthozoa, Phylum Cnidaria). *PLoS ONE* [Internet] 3(9): e3222. 9 p. Available at http://www.plosone.org/article/info:doi/10.1371/journal.pone.0003222

Furla, P., Galgani, I., Durand, I., Allemand, D. 2000. Sources and mechanisms of inorganic carbon transport for coral calcification and photosynthesis. *Journal of Experimental Biology* 203: 3445–457.

Gardner, T. A., Côté, I. M., Gill, J. A., Grant, A., Watkinson, A. R. 2003. Long-term region-wide declines in Caribbean corals. *Science* 301: 958–60.

Garzón-Ferreira, J., Zea, S., Díaz, J. M. 2005. Incidence of partial mortality and other health indicators in hard-coral communities of four southwestern Caribbean atolls. *Bulletin of Marine Science* 76(1): 105–22.

Gattuso, J. -P., Allemand, D., Frankignoulle, M. 1999. Photosynthesis and calcification at cellular, organismal and community levels in coral reefs: a review on interactions and control by carbonate chemistry. *American Zoologist* 39: 160–83.

Goldberg, W. M. 2002. Feeding behavior, epidermal structure and mucus cytochemistry of the scleractinian *Mycetophyllia reesi*, a coral without tentacles. *Tissue and Cell* 34(4): 232–45.

Goreau, T. F., Goreau, N. I. 1973. The ecology of Jamaican coral reefs. II. Geomorphology, zonation and sedimentary phases. *Bulletin of Marine Science* 23: 399–464.

Hatcher, B. G. 1997. Organic production and decomposition. In: Birkeland, C., editor. 1997. *Life and death of coral reefs*. New York: Chapman & Hall. p 140–74.

Hawkins, J. P., Roberts, C. M. 2004. Effects of artisanal fishing on Caribbean coral reefs. *Conservation Biology* 18: 215–26.

Hoegh-Guldberg, O., Mumby, P. J., Hooten, A. J., Steneck, R. S., Greenfield, P., Gomez, E., Harvell, C. D., Sale, P. F., Edwards, A. J., Caldeira, K., et al. 2007. Coral reefs under rapid climate change and ocean acidification. *Science* 318: 1737–42.

Huang, D., Licuanan, W. Y., Baird, A. H., Fukami, H. 2011. Cleaning up the 'Bigmessidae': molecular phylogeny of scleractinian corals from Faviidae, Merulinidae, Pectiniidae and Trachyphylliidae. *BMC Evolutionary Biology* [Internet] 11(37). 13 p. Available at http://www.biomedcentral.com/1471-2148/11/37

Humann, P., Deloach, N. 2002. *Reef coral identification: Florida Caribbean Bahamas.* 2nd ed. Jacksonville (FL): New World Publications. 278 p.

ISRS. 2007. Coral reefs and ocean acidification. Briefing Paper 5, International Society for Reef Studies [Internet]. 8 p. Available at http://www.coralreefs .org/isrs-briefing-papers-and-statements.htm

Kitahara, M. V., Cairns, S. D., Stolarski, J., Blair, D., Miller, D. J. 2010. A comprehensive phylogenetic analysis of the Scleractinia (Cnidaria, Anthozoa) based on mitochondrial CO1 sequence data. *PLoS ONE* [Internet] 5(7): e11490. 9 p. Available at http://www.plosone.org/article/ info%3Adoi%2F10.1371%2Fjournal.pone.0011490

Lang, J. C., editor. 2003. Status of coral reefs in the Western Atlantic: results of initial surveys, Atlantic and Gulf Rapid Reef Assessment (AGRRA) program. *Atoll Research Bulletin* 496. 625 p.

Lang, J. C., Chornesky, E. A. 1990. Competition between scleractinian reef corals: a review of mechanisms and effects. In: Dubinsky, Z., editor. *Coral reefs.* Ecosystems of the World 25. Amsterdam: Elsevier. p 209–52.

Lessios, H. A., Robertson, D. R., Cubit, J. D. 1984. Spread of *Diadema* mass mortality through the Caribbean. *Science* 226: 335–37.

López-Victoria, M., Zea, S., Weil, E. 2006. Competition for space between encrusting excavating Caribbean sponges and other coral reef organisms. *Marine Ecology Progress Series* 312: 113–21.

Marks, K. W. 2007. Atlantic and Gulf Rapid Reef Assessment (AGRRA) Database. Version (10/2007) [Internet]. Available at http:www.agrra.org/ Release_2007-10/

Mendes, J. 2004. Timing of skeletal band formation in *Montastraea annularis*: relationship to environmental and endogenous factors. *Bulletin of Marine Science* 75(3): 423–37.

Mendes, J. M., Woodley, J. D. 2002. Effect of the 1995–1996 bleaching event on polyp tissue depth, growth, reproduction and skeletal band formation in *Montastraea annularis. Marine Ecology Progress Series* 235: 93–102.

Mora, C. 2008. A clear human footprint in the coral reefs of the Caribbean. *Proceedings of the Royal Society B* 275: 767–73.

O'Dea, A., Jackson, J.B.C., Fortunato, H., Smith, J. T., D'Croz, L., Johnson, K. G., Todd, J. A. 2007. Environmental change preceded Caribbean extinction by 2 million years. *Proceedings of the National Academy of Sciences* 104: 5501–6.

Richardson, L. L., Miller, A. W., Broderick, E., Kaczmarsky, L., Gantar, M., Stanić, D., Sekar, R. 2009. Sulfide, microcystin, and the etiology of black band disease. *Diseases of Aquatic Organisms* 87(1–2): 79–90.

Richmond, R. H. 1997. Reproduction and recruitment in corals: critical links in the persistence of reefs. In: Birkeland, C., editor. 1997. *Life and death of coral reefs*. New York: Chapman & Hall. p 175–97.

Richmond, R. H., Hunter, C. L. 1990. Reproduction and recruitment of corals: comparisons amongst the Caribbean, the tropical Pacific and the Red Sea. *Marine Ecology Progress Series* 60: 185–203.

Ruiz Ramos, D. V. 2009. Morphological and genetic variation in the Caribbean species of the hydrocoral genus *Millepora* [thesis]. Mayagüez: University of Puerto Rico. 77 p.

Schutte, V.G.W., Selig, E. R., Bruno, J. F. 2010. Regional spatio-temporal trends in Caribbean coral reef benthic communities. *Marine Ecology Progress Series* 402: 115–22.

Stake, J. L. 2007. Novel molecular markers for phylogenetic studies of scleractinian corals [dissertation]. Lafayette: University of Louisiana. 165 p.

Stanley, G. D. Jr. 2003. The evolution of modern corals and their early history. *Earth-Science Reviews* 60: 195–225.

Sutherland, K. P., Porter, J. W., Torres, C. 2004. Disease and immunity in Caribbean and Indo-Pacific zooxanthellate corals. *Marine Ecology Progress Series* 266: 273–302.

Szmant, A. M. 2002. Nutrient enrichment on coral reefs: is it a major cause of coral reef decline? *Estuaries* 25: 743–66.

Vermeij, M.J.A., Bakker, J., van der Hal, N., Bak, R.P.M. 2011. Juvenile coral abundance has decreased by more than 50% in only three decades on a small Caribbean island. *Diversity* 3: 296–307.

Vermeij, M.J.A., Diekmann, O. E., Bak, R.P.M. 2003. A new species of scleractinian coral (Cnidaria, Anthozoa), *Madracis carmabi* n. sp. from the Caribbean. *Bulletin of Marine Science* 73(3): 679–84.

Veron, J.E.N. 2000. *Corals of the world*. 1–3. Townsville, Australia: Australian Institute of Marine Sciences. 463, 429, 490 p.

Veron, J.E.N. 2008. Mass extinctions and ocean acidification: biological constraints on geological dilemmas. *Coral Reefs* 27: 459–72.

Vollmer, S. V., Kline, D. I. 2008. Natural disease resistance in threatened staghorn corals. *PLoS ONE* [Internet] 3(11): e3718. 5 p. Available at http://www.plosone.org/article/info:doi%2F10.1371%2Fjournal.pone.0003718

Weis, V. M. 2008. Cellular mechanisms of cnidarian bleaching: stress causes the collapse of symbiosis. *Journal of Experimental Biology* 211: 3059–66.

Wells, J. W. 1973. New and old scleractinian corals from Jamaica. *Bulletin of Marine Science* 23: 16–58.

Wilkinson, C., editor. 2008. *Status of coral reefs of the world: 2008.* Townsville, Australia: Global Coral Reef Monitoring Network and Rainforest Research Center. 296 p.

# Index

George F. Warner is a retired marine ecologist, recently employed as director of the Centre for Marine Sciences at the University of the West Indies, Kingston, Jamaica, and as consultant at the University of Trinidad and Tobago at the Institute of Marine Affairs, Chaguaramas, Trinidad. Most of his career was spent at the University of Reading, United Kingdom, the country where he now lives.